The Blindspot Before the Disaster

Major organisational disasters are not random. They are predictable. In the months and years before every catastrophic failure, before the Deepwater Horizon blowout, the Grenfell Tower fire, the Space Shuttle disasters, the Pike River explosion, the warning signs were already present. They were systematically unseen. Organisational blindness to accumulating risk is not a malfunction. In many organisations, it is the system working exactly as designed. This book explains why and what can be done about it.

The Blindspot Before the Disaster identifies the five organisational conditions that converge to produce this failure of sight: Signal Attenuation, Normalised Tolerance, Structural Secrecy, Production Gravity, and Leadership Detachment. Together they form the Blindspot Model, a practical diagnostic framework grounded in four decades of disaster scholarship and demonstrated across 11 of the most significant industrial catastrophes of the past 40 years.

This book is written for the leaders, safety professionals, and governance practitioners who are responsible for preventing harm before it occurs. It bridges the gap between rigorous academic disaster theory and the practical needs of people working in high-consequence environments. It does not assign blame. It offers something more useful: a structured, evidence-based way of seeing the danger that is already in the room.

Tom Bourne is an Australian author, workplace safety educator, speaker, and leadership commentator with more than 25 years of experience across the mining and resources, government, and professional education sectors. He has published more than 16 books on workplace safety, leadership, and organisational accountability. He is based in Perth, Western Australia.

The Blindspot Before the Disaster

A Safety Model for Developing Organisational Foresight

Tom Bourne

Routledge
Taylor & Francis Group
LONDON AND NEW YORK

Designed cover image: Getty Images. Credit: TomasSereda

First published 2027
by Routledge
4 Park Square, Milton Park, Abingdon, Oxon OX14 4RN

and by Routledge
605 Third Avenue, New York, NY 10158

Routledge is an imprint of the Taylor & Francis Group, an informa business

British Library Cataloguing-in-Publication Data
A catalogue record for this book is available from the British Library

ISBN: 978-1-041-39343-6 (hbk)
ISBN: 978-1-041-39337-5 (pbk)
ISBN: 978-1-042-01753-9 (ebk)

DOI: 10.4324/9781042017539

Typeset in Sabon
by codeMantra

Contents

About the Author

Tom Bourne is an Australian author, workplace safety educator, leadership commentator, and trainer working at the intersection of safety science, organisational behaviour, and ethical leadership. His work is grounded in the conviction that understanding why organisations fail is more useful and more honest than assigning blame to the individuals who happened to be present when systems collapsed, and that developing the organisational capacity to see accumulating danger is the most important and most underinvested challenge in workplace safety today.

Tom's writing spans the analysis of major industrial disasters, the ethics of leadership accountability, the practical application of risk management frameworks, and the cultural and organisational conditions that produce workplace harm. He brings to each of these areas a combination of rigorous engagement with the academic and regulatory literature and the direct, practical experience of working with leaders and organisations in Australian high-hazard industries.

His published books include the Ethics Trilogy, *The Architecture of Blame*, *The Courage Contract*, and *The Flying Monkey Field Guide*, which examine the relationship between organisational culture, ethical decision-making, and the accountability frameworks through which leaders manage the consequences of organisational failure. His book *Reasonably Practicable* provides Australian leaders and safety professionals with a practical guide to understanding how courts reason about workplace harm and what the legal standard of reasonably practicable actually requires in operational practice. His research into the Wittenoom asbestos disaster represents one of the most detailed examinations available of the corporate and regulatory failure that produced Australia's worst industrial public health catastrophe.

The Blindspot Before the Disaster is Tom's most ambitious and most intellectually substantial work. It draws together his research into organisational disaster, his analysis of the safety science literature, and his direct professional experience to produce an original analytical framework, the Blindspot Model, that synthesises four decades of disaster scholarship into a single, practically oriented account of how organisations develop the conditions for catastrophic harm without seeing them developing.

Tom is the host of the *Health and Safety Conversations* podcast and the *Voices from the Edge* podcast platform, through which he engages with safety professionals, leaders, investigators, and researchers from across the Australian and international safety community. These platforms reflect his commitment to the translation of safety science from the academic literature into the practical understanding of the practitioners who need it most.

He is a respected commentator on workplace safety, leadership accountability, and organisational ethics, whose work appears in professional publications, conference proceedings, and public safety discourse across Australia and internationally. He is a sought-after speaker and facilitator for organisational leadership programmes, safety leadership development initiatives, and incident investigation training.

Tom works primarily in Western Australia, where the mining, oil and gas, and construction industries that are central to the state's economy provide a constant and demanding context for the application of the ideas this book develops. He brings to that work the analytical rigour that the evidence demands and the directness that the people he works with deserve.

tombourne.com.au
Health and Safety Conversations Podcast
Voices from the Edge Podcast

Preface: Why I Wrote This Book

Why I Wrote This Book

There is a particular kind of silence that falls over a room when an experienced safety professional admits, quietly and without theatrics, that they saw it coming.

I have been in those rooms. I have sat across tables from people who knew, not with the clarity of hindsight, but with the uncomfortable, half-formed certainty of someone who has spent years watching how organisations work, that something was wrong. That the pressure was too high, or the maintenance too deferred, or the warnings too consistently reframed as acceptable variations. They knew, in the way that people embedded in complex systems learn to know things, that the organisation was moving in a dangerous direction. And they did not stop it. Not because they were cowards or because they did not care, but because the machinery of organisational life, the structures, the cultures, the hierarchies, the economics, made it extraordinarily difficult to translate that knowledge into action.

And then something happened. And the investigations began. And the reports were written. And the individuals who had been present at the critical moments were named, their decisions dissected, their failures catalogued. And the organisations moved on, having learned the lessons, having implemented the recommendations, having assured themselves and their regulators that this would not happen again.

And then, in another organisation, often in another industry, under different leadership and in a different regulatory jurisdiction, it happened again.

This book is my attempt to understand why.

I have spent years working in and around the Australian safety and leadership environment. I have investigated incidents, written about disasters, taught risk frameworks, and sat with leaders who were genuinely committed to keeping their people safe and yet who presided over organisations that were quietly, incrementally, invisibly accumulating the conditions for catastrophic harm. I have also spent years reading the scholarship on organisational disaster, the foundational theoretical work of Barry Turner, the forensic case analysis of Diane Vaughan, the applied intelligence of Andrew Hopkins, and the conceptual reframings of Sidney Dekker. It is extraordinary work, produced by extraordinary minds, and it has advanced our understanding of how disasters happen enormously.

What I could not find was a book that took all of that scholarship and answered the question I kept being asked by the people I worked with. Not the academic question of why disasters occur, the literature addresses that with considerable rigour. But the practical question that follows from it: given everything we know about why organisations become blind to accumulating danger, what can we actually do about it before someone gets killed?

That is the question this book is written to answer.

The argument I make is not a comfortable one. It asks organisations and their leaders to accept that the conditions producing catastrophic harm are not aberrations, not the consequences of uniquely bad decisions made by uniquely bad people in uniquely dysfunctional organisations. They are the predictable outputs of ordinary organisational mechanisms operating under ordinary pressures. The Blindspot is not a malfunction. In many ways, it is the system working exactly as designed. That is what makes it so difficult to see, and so dangerous to ignore.

I have tried to write a book that is intellectually honest about the complexity of these problems while remaining practically useful to the people who have to navigate them. The academic literature I draw on is extensive and, where possible, I have tried to represent it faithfully. Where I have developed my own framework and my own arguments, I have tried to be clear about where the evidence supports them and where reasonable people might disagree.

The disasters I write about in this book involved real people: workers, families, communities, and the leaders and organisations whose decisions shaped their fate. I have tried to treat those stories with the seriousness they deserve, neither dramatising them for effect nor reducing them to analytical abstractions. They are not case studies in the detached sense. They are what happens when organisations lose the ability to see the danger that was already in front of them. Understanding that, really understanding it, seems to me one of the most important things any person responsible for the safety of others can do.

This book is for those people.

Introduction

The Problem of Seeing

The Problem of Seeing

Two Disasters, One Pattern

On the morning of 23 January 1998, workers at the Esso gas processing plant at Longford in Victoria's Gippsland region were managing a routine operation involving heat exchange equipment that had been running for years without significant incident. The processes they were following were familiar. The plant had operated continuously since 1969. The people on shift that morning were experienced. And yet, in the space of a few hours, a sequence of events unfolded that no one at the plant recognised as catastrophic until it was too late to prevent it. An explosion and fire killed two workers, injured eight others, and cut off gas supply to the state of Victoria for two weeks, the longest and most economically significant industrial disruption in Australia's history to that point.

The subsequent Royal Commission, conducted by Justice Dawson and Professor Brooks, found that the accident was neither random nor unforeseeable. The conditions that produced it had been developing for years. The engineering hazard at the heart of the disaster, the behaviour of heat exchanger equipment under certain abnormal temperature conditions, was known, had been studied, and had been documented. The operational practices that contributed to the accident had deviated progressively from those originally designed for the plant. The training systems that might have identified those deviations had deteriorated. The communication structures between the operating company and its parent corporation had, over time, produced exactly the kind of fragmentation of safety-critical knowledge that the disaster literature would later identify as a defining feature of organisational catastrophe (Hopkins, 2000).

Twelve years later, on 20 April 2010, workers on the Deepwater Horizon drilling platform in the Gulf of Mexico were preparing to temporarily abandon the Macondo well in preparation for a handover. The operation was behind schedule. Commercial pressures were significant. A series of decisions

DOI: 10.4324/9781042017539-1

had been made in the preceding days that, taken individually, each appeared defensible within the operational and economic context of deepwater drilling. Taken together, they had eliminated the safety margins separating the well from an uncontrolled blowout. At 9:49 that evening, gas and fluid from the well breached the platform. The explosion that followed killed 11 workers. The resulting oil spill, the largest accidental marine oil spill in history, caused environmental and economic damage that continues to be felt today (Hopkins, 2012; National Commission on the BP Deepwater Horizon Oil Spill and Offshore Drilling, 2011).

The subsequent investigations – the National Commission report, the Bureau of Safety and Environmental Enforcement inquiry, and Hopkins' own analysis in *Disastrous Decisions* – found, again, that the conditions for the disaster had been developing over an extended period. Warning signs had been present. Decisions that eroded safety margins had accumulated incrementally. The individuals involved were not reckless or malicious. They were operating within the ordinary constraints and incentives of their organisations, making decisions that the structures around them made logical, and failing to see the danger that those decisions, in combination, were producing.

Different countries. Different industries. Different technologies. Different organisations and different people. The same pattern.

The most dangerous period in any organisation's safety history is not when something goes catastrophically wrong. It is the period before that moment, when the evidence of failure is already present in the system, and yet consistently unseen by the people responsible for preventing harm.

What the Evidence Shows

The observation that major industrial disasters are preceded by periods of accumulating, unrecognised warning signals is not new. Barry Turner first articulated it with formal rigour in *Man-Made Disasters*, originally published in 1978 and revised with Nick Pidgeon in 1997. Turner argued that disasters do not erupt from nowhere; they incubate. In the period preceding a major organisational failure, information relevant to the hazard exists within the system. It is present in incident reports and near-miss logs, in the informal knowledge of experienced operators, in the observations of people working closest to the hazard. What distinguishes disasters from near-misses is not the absence of that information but the failure of the organisation to register it, interpret it correctly, and act on it in time (Turner and Pidgeon, 1997).

Turner called this the incubation period, and its identification was a foundational contribution to the sociology of disaster. It shifted the focus of disaster analysis from the precipitating event – the explosion, the capsize, the blowout – to the extended organisational history that made that event possible. It asked a different set of questions: Not just what happened, but why the conditions that allowed it to happen were able to develop over time without being recognised and addressed.

In the decades since Turner's original formulation, a substantial body of scholarship has elaborated and refined this foundational insight. Diane Vaughan's forensic analysis of the Space Shuttle Challenger disaster demonstrated how the process she called normalisation of deviance operated within NASA, how repeated exposure to a known technical anomaly in the O-ring seals gradually transformed a recognised engineering concern into an accepted characteristic of shuttle operations, until the decision to launch in conditions known to be outside the seals' certified temperature range appeared, to those making it, not as a dangerous deviation from established safety practice but as a reasonable operational judgement (Vaughan, 1996). Charles Perrow's concept of normal accidents argued that certain categories of complex, tightly coupled technological systems are, by their nature, prone to failures that originate in the routine interactions of their components and that cannot be prevented by improving individual component performance alone (Perrow, 1999). James Reason's organisational accident model demonstrated how latent conditions – pre-existing weaknesses embedded in organisational structure, culture, and management practice – align with local triggers to produce catastrophic outcomes (Reason, 1997). Sidney Dekker reframed the entire conversation by arguing that organisations do not drift into failure because they are malfunctioning, they drift because the system is working as it was designed to work, optimising for production and efficiency in ways that make incremental risk-taking not just possible but rational (Dekker, 2011).

Each of these contributions advanced the field. Together, they constitute a body of evidence that is, in the relevant scholarly communities, essentially settled: major organisational disasters are, in the general case, not random events caused by isolated failures or individual incompetence. They are the predictable outcomes of identifiable organisational conditions that develop over time and that are, in principle, detectable and preventable.

The question this book addresses is why, given this evidence, organisations continue to produce those conditions at a rate that shows no signs of meaningful decline.

The Limits of the Existing Literature

The disaster literature is rich, but it has a structural limitation that this book is designed to address.

The foundational theoretical work – Turner and Pidgeon, Reason, Perrow, and Weick – is powerful and important, but it is primarily written for an academic audience. It is conceptually sophisticated, theoretically dense, and often explicitly concerned with advancing scholarly debate rather than providing practical guidance to the people responsible for preventing disasters in real organisations. This is not a criticism. It is an accurate description of what that work is for, and it does it well. But the leaders, safety professionals, and investigators who need to apply these insights in operational environments cannot always traverse the gap between theoretical frameworks and practical application without substantial support.

The applied disaster literature – Andrew Hopkins' series of case analyses, Rebecca Macfie's account of Pike River, and Adam Higginbotham's narrative histories of Challenger and Chernobyl – is more accessible and directly applicable. Hopkins in particular has demonstrated, across more than two decades of meticulous disaster analysis, how the insights of disaster theory can be applied to specific cases in ways that yield concrete, actionable findings (Hopkins, 2000, 2007, 2008, 2012, 2019). But the case-study format, by its nature, organises understanding around individual disasters rather than around the underlying mechanisms those disasters share. It produces deep understanding of specific events without necessarily generating a generalisable framework that practitioners can apply prospectively, not just retrospectively, to understand what went wrong, but prospectively, to identify what is going wrong before it becomes irreversible.

What is missing from the literature is a book that does all of the following simultaneously: grounds itself rigorously in the academic scholarship; synthesises that scholarship into a single, coherent, named framework that practitioners can learn and apply; demonstrates that framework's operation across a range of cases in a structured and consistent way; and provides practical, evidence-based guidance for the leaders and safety professionals who want to develop their organisations' capacity to see accumulating danger before it produces catastrophic harm.

That is what this book attempts to do.

The Blindspot Model

The central analytical contribution of this book is what I call the Blindspot Model: a framework that identifies five interlocking organisational conditions which, when operating together, reliably produce the inability to see accumulating danger in time to prevent catastrophic harm.

The five conditions are:

Signal Attenuation, the systematic filtering, softening, and loss of safety-critical information as it moves upward through organisational hierarchies, such that warning signs present at the operational level fail to reach the decision-makers with the authority and responsibility to act on them.

Normalised Tolerance, the gradual, culturally embedded process through which repeated exposure to a risk that does not immediately produce harm generates a revised, more permissive understanding of what constitutes acceptable operational practice, until conditions that were once recognised as hazardous are treated as routine.

Structural Secrecy, the fragmentation of safety-critical information across organisational boundaries, functional silos, geographic sites, and hierarchical layers, such that no individual or team has access to the complete picture of accumulating risk, even when that picture, assembled, would be clearly alarming.

Production Gravity, the constant economic and operational pressure that pulls organisational decision-making toward the performance of productive work and away from the maintenance of safety margins, producing incremental, individually justified erosions of those margins that, in combination, leave the organisation without adequate protection against catastrophic failure.

Leadership Detachment, the structural loss of contact between senior leaders and operational reality that is a predictable consequence of organisational scale, hierarchy, and information filtering, producing a leadership team whose confidence in the organisation's safety performance is grounded not in accurate knowledge of the system's actual state but in the reassuring picture that the system's information flows have constructed for them.

None of these conditions is, on its own, new to the literature. Signal Attenuation has been discussed under various names since Turner's original work on incubation. Normalised Tolerance is Vaughan's normalisation of deviance, one of the most replicated findings in the sociology of risk. Structural secrecy is Vaughan's own term, developed in *The Challenger Launch Decision*. Production Gravity is closely related to Dekker's account of drift, and to the concept of practical drift developed by Scott Snook in his analysis of the Black Hawk shootdown. Leadership Detachment draws on Hopkins' work on mindful leadership and on Weick and Sutcliffe's analysis of high-reliability organisations.

What is new is the synthesis: the identification of these five conditions as a coherent, interacting set; the argument that their combination is what produces the most severe and most consistent forms of organisational blindness; and the framework this combination provides for prospective assessment of organisational risk states rather than retrospective analysis of disaster causes.

The Cases

Each of the five conditions is examined in depth through detailed analysis of major international disasters and incidents. The cases selected are not chosen for dramatic effect, though some of them are among the most dramatic events of the past half-century. They are chosen because the documentary record, in the form of Royal Commission reports, independent inquiry findings, academic analyses, and first-hand accounts, is sufficiently detailed and sufficiently credible to support the kind of close analytical reading that the Blindspot Model requires.

The primary cases examined in this book are: the Deepwater Horizon blowout (Gulf of Mexico, 2010); the Grenfell Tower fire (London, 2017);

the Space Shuttle Challenger disaster (1986); the Space Shuttle Columbia disaster (2003); the Pike River mine disaster (New Zealand, 2010); the Esso Longford gas plant explosion (Victoria, Australia, 1998); the BP Texas City refinery explosion (Texas, USA, 2005); the Wittenoom asbestos disaster (Western Australia, 1950s to 1966); the Chernobyl nuclear disaster (Soviet Union, 1986); the Gretley mine disaster (New South Wales, Australia, 1996); and the capsize of the Herald of Free Enterprise (off Zeebrugge, Belgium, 1987).

Together, these cases span multiple industries, multiple countries, multiple regulatory environments, and multiple decades. The consistency of the patterns they reveal is itself a significant finding. Organisational blindness to accumulating risk is not an artefact of a particular industry, a particular regulatory culture, or a particular era. It is a general feature of how large organisations operating under production pressure manage, or fail to manage, safety-critical information and decision-making.

What This Book Is Not

A brief note on what this book does not attempt to do, because clarity on this point will help the reader understand what it does.

This is not a book that assigns blame. The case analyses presented here are not intended to identify villains or to relitigate the findings of inquiries and Royal Commissions. In most of the cases examined in this book, the individuals who made the decisions that contributed to disaster were not malicious, not incompetent, and not indifferent to the welfare of the people in their care. They were people operating within organisational systems that produced predictable patterns of blindness, systems that, in many cases, would have produced similar outcomes regardless of which specific individuals occupied the relevant roles. Understanding that does not absolve individuals of accountability. It does, however, shift the focus of analysis to where it is most productive: the organisational conditions that made those individual failures likely, rather than the individual failures themselves.

This is also not a book that argues safety is impossible in high-hazard industries, or that complex sociotechnical systems are inherently beyond our ability to manage safely. The evidence does not support that conclusion. The final part of this book examines organisations and leadership teams that have successfully developed the capacity to see accumulating danger before it becomes catastrophic harm, and what they did to build that capacity. The challenge is real and serious. It is not insurmountable.

Finally, this is not a book for academics, although it draws extensively on academic scholarship, and I hope academics will find it useful. It is written for the people who are professionally responsible for the safety of others and who want a deeper, more honest, more evidence-based understanding of why the organisations they work in and lead sometimes fail to keep people safe, and what can be done about it.

How to Read This Book

This book is organised in four parts.

Part I, *The Anatomy of Blindness*, establishes the intellectual foundations. It examines the dominant mental models of disaster causation that shape how organisations respond to incidents and why those models are inadequate; it introduces Turner's concept of incubation in detail; and it presents the Blindspot Model in full.

Part II, *The Five Conditions*, examines each of the five conditions of the Blindspot Model in depth, using detailed case-study analysis to demonstrate how each condition operates in practice, what its characteristic indicators are, and what distinguishes organisations where it is present from those where it has been successfully managed.

Part III, *When the Conditions Combine*, examines two cases in which all five conditions operated simultaneously, demonstrating both the explanatory power of the Blindspot Model and the particular danger of organisational states in which no single condition is providing any compensating protection. It also examines the ethical dimension of organisational blindness: the relationship between the Blindspot Model and the moral accountability of organisations and their leaders.

Part IV, *Active Foresight*, turns from analysis to application. It examines the evidence from organisations that have successfully interrupted the Blindspot pattern, and it provides a structured practical framework for leaders and safety professionals who want to assess and address the five conditions in their own organisations.

Each chapter ends with a summary of its key arguments and a chapter reference list. A full bibliography appears at the end of this book, along with a glossary of key terms and concepts.

The reader who wants the practical framework without the full theoretical grounding can proceed directly to Part IV after reading Part I. The reader who wants to understand the evidence base as deeply as possible should read this book in sequence. Both approaches will, I hope, be rewarded.

What I ask of every reader is this: approach the evidence with the same rigour and honesty that the people responsible for safety in your organisation deserve. The disasters described in this book were not inevitable. They were the products of choices, not always the dramatic, visible choices that appear in inquiry reports, but the quiet, incremental, individually reasonable choices that organisations make every day, and that collectively determine whether they can see the danger in front of them before it is too late to act.

That capacity to see, to genuinely, honestly, and systematically see, is what this book is about.

Part I

The Anatomy of Blindness

The Anatomy of Blindness

Chapters 1–3

The question is not why disasters happen. After five decades of serious scholarship, we have a reasonably clear answer to that. The question is why organisations that know the answer continue to produce the conditions that make disasters inevitable.

DOI: 10.4324/9781042017539-2

Chapter 1

What We Think Causes Disasters, and Why We Are Wrong

The dominant mental model of organisational failure, why it persists, and why the evidence demands something better.

The Story We Tell

Every major industrial disaster generates a story. Not the full story, the one with all its complexity, its institutional history, its accumulating pressures, its years of small decisions, but a story. The story is produced by the inquiry process, shaped by the legal and political environment in which that process operates, and then transmitted to the public through media coverage that further simplifies and condenses it. By the time it reaches most people, including most of the people who work in high-hazard industries and who are, ostensibly, learning from it, the story has a standard structure.

There was an event. People died or were harmed. An investigation was conducted. Failures were identified. Those failures were attributed, in most cases, to a combination of individual error, inadequate procedures, and, where the evidence was particularly compelling, a failure of management to maintain appropriate standards. Recommendations were made. The organisation reformed. The lessons were learned.

This story is not always entirely false. Individuals do make errors. Procedures are sometimes inadequate. Management does sometimes fail to maintain standards. The reforms and recommendations that follow inquiries are not always meaningless, and some of them genuinely improve safety outcomes. But as a model for understanding why organisations produce catastrophic harm, and, more importantly, as a guide to preventing them from doing so again, the standard story is seriously and consistently inadequate.

It is inadequate not because it is dishonest, but because it is structurally limited. It is shaped by processes – legal, political, institutional – that are designed to answer particular questions: who was responsible, what rules were broken, what standards were not met? These are legitimate questions, and the answers to them have real consequences for accountability and for

DOI: 10.4324/9781042017539-3

the regulatory framework. But they are not the same questions as: why did this organisation develop the conditions that made this disaster possible, and how can organisations be prevented from developing those conditions in the future? Those questions require a different kind of analysis, and the dominant story does not provide it.

The result is a pattern that has repeated itself with remarkable consistency across five decades of industrial disaster. An event occurs. An inquiry is conducted. A story is produced. Lessons are declared to have been learned. And then, in another organisation, the same conditions develop, the same blindness sets in, and the same outcome eventually results.

> *If the lessons of one disaster were genuinely being learned in the form in which the evidence requires, the same patterns would not recur. Their recurrence is evidence that something is wrong not with the organisations that produce disasters, but with the framework through which we try to understand them.*

The Blame Attribution Model

The dominant mental model of disaster causation, the one that shapes most inquiry processes, most regulatory responses, and most organisational learning efforts, can be described as the blame attribution model. Its defining feature is the identification of discrete failures at identifiable points in a causal chain, and the attribution of those failures to specific individuals or specific systems that did not perform as required.

The blame attribution model has several characteristics that make it politically and institutionally attractive. It provides clear answers to the questions that legal and regulatory processes require to be answered. It produces identifiable accountable parties. It generates specific, actionable recommendations – retrain this person, fix this procedure, enforce this standard – that organisations can implement and regulators can verify. It satisfies, at least temporarily, the public and institutional demand for accountability that follows major disasters.

It also has several characteristics that make it analytically inadequate. The most important of these is that it treats disasters as chains of discrete failures rather than as the outputs of complex organisational systems, and it treats those failures as aberrations, deviations from a normally functioning system, rather than as the predictable products of how that system actually operates.

Dekker (2002) has described this as the old view of human error: the tendency to treat human failure as the cause of accidents rather than as a symptom of deeper system problems. The old view, he argues, starts from the assumption that systems are basically safe and that accidents happen when humans deviate from expected performance. The new view, grounded in decades of human factors research, organisational psychology, and disaster sociology, starts from the assumption that complex systems contain latent conditions that make certain kinds of failure not just possible but, under the right circumstances, likely, and that human behaviour makes sense in the context of the local conditions in which it occurs.

The transition from the old view to the new is not merely an academic distinction. It has profound practical implications for how organisations respond to disasters and, more importantly, for whether those responses actually reduce the probability of future harm. An organisation that responds to a disaster by identifying and disciplining the individuals who made the critical errors, without examining the systemic conditions that shaped those decisions, has not learned the lesson. It has performed the appearance of learning while leaving the conditions for the next disaster intact.

The Inquiry Process and Its Limitations

The limitations of the blame attribution model are reinforced by the structural characteristics of the inquiry processes through which disaster analysis is most commonly conducted.

Most formal inquiries into major industrial disasters operate under at least some degree of legal constraint. The individuals who appear before them may face criminal prosecution, civil liability, or professional sanction as a result of their findings. This creates powerful pressures, both on the individuals giving evidence and on the inquiry process itself, toward the framing of causation in terms of individual error and procedural non-compliance rather than in terms of systemic organisational conditions. Legal accountability requires identifiable individual actors. Systemic analysis does not map neatly onto legal frameworks, and inquiry processes that operate within those frameworks are constrained accordingly.

Hopkins (2008) has noted that the inquiries following major industrial disasters tend to focus on what he calls proximate causes, the immediate, visible triggers of the disaster event, at the expense of the organisational and management causes that created the conditions making those proximate causes dangerous. This is not a universal feature of inquiry processes. Some of the most valuable analyses of major disasters have come from inquiries that explicitly looked beyond proximate causation – the Longford Royal Commission, the Columbia Accident Investigation Board, the Pike River Royal Commission among them. But the tendency is real, and its consequences for organisational learning are significant.

The Columbia Accident Investigation Board (2003) made this point explicitly in its report on the Space Shuttle Columbia disaster, arguing that NASA's organisational culture and decision-making processes were as much a cause of the disaster as the physical failure of the thermal protection system.

The Board observed that NASA had not learned the lessons of the Challenger disaster 17 years earlier, not because the lessons were not available, but because the organisational conditions that produced Challenger had not been fundamentally altered by the response to it. The same normalisation of deviance, the same production pressure, the same structural barriers to the upward transmission of safety concerns, had simply reasserted themselves in a new operational context.

> *The most dangerous conclusion an organisation can draw from a disaster investigation is that the problem has been fixed. The problem is rarely the specific failure that the investigation identifies. The problem is the organisational conditions that made that failure possible, and those conditions do not disappear because a procedure has been rewritten or a supervisor has been retrained.*

What the Evidence Actually Shows

The evidence accumulated by 50 years of disaster scholarship tells a different story from the one produced by most inquiry processes. It is a more complex story, and in some respects a more uncomfortable one, because it does not yield the kind of clear individual accountability that inquiry processes are designed to produce. But it is a more accurate story, and that accuracy is what makes it useful for the purpose of actually preventing disasters rather than simply responding to them.

The story the evidence tells begins with Turner's foundational insight: disasters incubate. In the period preceding a major organisational failure, the conditions for that failure develop gradually, accumulating over months or years, generating warning signals that are present within the system but that are not correctly registered, interpreted, or acted upon by the organisation (Turner and Pidgeon, 1997). The precipitating event, the explosion, the blowout, the structural failure, is not the beginning of the disaster. It is the end of a long process of organisational development that produced the conditions making the precipitating event catastrophic rather than merely unfortunate.

This insight was elaborated and refined by Vaughan's (1996) analysis of the Challenger disaster, which demonstrated with forensic precision how the process of normalisation of deviance operated within NASA over a period of years, gradually transforming a recognised engineering concern into an accepted operational characteristic. It was extended by Reason's (1997) organisational accident model, which showed how latent conditions embedded in organisational structure and management practice align with active failures at the sharp end of operations to produce catastrophic outcomes. It was reframed by Dekker (2011), who argued that organisations drift into failure not through dysfunction but through the ordinary operation of systems optimising for production and efficiency under conditions of incomplete information and competitive pressure.

What these analyses share, and what distinguishes them from the blame attribution model, is their focus on the organisational conditions that precede disaster rather than on the individual decisions and failures that trigger it. They ask not who made the wrong decision, but what kind of organisation produces conditions in which wrong decisions become likely. And they demonstrate, consistently across a wide range of cases, that those conditions are not random or unique to particular organisations. They are patterned. They recur. They can be identified and, in principle, interrupted.

Reason (2000) captured this with his famous metaphor of the Swiss Cheese model: a system of defences against failure in which each layer of defence has gaps, and in which disasters occur when those gaps momentarily align, allowing a trajectory of accident opportunity to pass through all layers simultaneously. The metaphor is illuminating, but its most important implication is often missed. The gaps do not appear randomly. They are produced by the same organisational conditions – Production pressure, Normalised Tolerance, Structural secrecy, Leadership Detachment – that this book identifies as the components of the Blindspot. Understanding where the gaps come from, and why they develop where they do, is what transforms Reason's metaphor from a retrospective description of disaster causation into a prospective tool for disaster prevention.

Why the Standard Story Persists

If the blame attribution model is as inadequate as the evidence suggests, why does it persist? Why do inquiry processes continue to produce it, why do organisations continue to respond to it, and why does the pattern of recurrence continue despite 50 years of scholarship demonstrating that something more is required?

There are several answers to this question, and they are worth examining because understanding them is part of understanding the problem this book addresses.

The first is institutional. Inquiry processes are designed primarily to assign accountability, not to understand organisational systems. The legal and regulatory frameworks within which they operate require them to answer questions about individual responsibility, and those frameworks are not easily adapted to accommodate the kind of systemic analysis that the evidence demands. This is not a conspiracy or a deliberate evasion. It is the structural consequence of designing accountability processes around individual decision-making rather than organisational conditions.

The second is psychological. The attribution of causation to individual error is cognitively natural. Kahneman (2011) has demonstrated extensively that human beings are prone to what social psychologists call the fundamental attribution error: the tendency to explain others' behaviour in terms of their personal characteristics rather than the situational factors constraining them. Applied to disaster analysis, this tendency produces exactly the blame attribution model: the identification of the individuals who made the critical decisions, and the attribution of those decisions to their personal failures of competence, judgement, or commitment.

The third is organisational. The blame attribution model is, for organisations, more comfortable than its alternative. It produces an explanation that does not implicate the organisation's fundamental structures, cultures, and management practices. It generates corrective actions that are bounded, specific, and achievable. It allows the organisation to declare, with some credibility, that the lessons have been learned and the problem has been fixed. The systemic alternative, examining the organisational conditions that produced the disaster, is more threatening, more disruptive, and more expensive. It implicates leadership. It requires genuine cultural change rather than procedural adjustment. It takes longer and produces less certain outcomes.

The fourth is political. The organisations that experience major disasters, and the industries of which they are part, have significant resources and significant interest in the framing of the causal story that inquiry processes produce. Those resources are often deployed, through legal representation and institutional engagement with inquiry processes, in ways that tend to narrow the causal frame, emphasise individual decision-making, and resist the kind of systemic organisational analysis that would be more damaging to the organisation's reputation and legal position.

None of these factors is, on its own, determinative. And none of them operates through deliberate dishonesty in most cases. They are structural tendencies that shape, at the margin, the kind of causal analysis that disaster inquiry processes produce, and, through those processes, the kind of organisational learning that follows.

The Cost of Inadequate Learning

The cost of this inadequacy is not abstract. It is measured in lives.

Jessie Singer's (2022) examination of accidents and disasters in the United States documents with considerable rigour the extent to which the framing of harm as the product of individual error or negligence, rather than as the product of systemic organisational and social conditions, has shaped public policy, regulatory practice, and corporate behaviour in ways that consistently prioritise the interests of organisations over the safety of the people who work in and around them. Singer's argument is, in important respects, broader than the one this book makes. But her core observation applies with particular force to the disaster context: when we attribute harm to individual failure, we exempt the systems that produce that harm from the scrutiny and reform they require.

Hopkins (2008) documented the aftermath of the BP Texas City refinery disaster of 2005, in which 15 workers were killed and 180 were injured, and which was preceded by an extended period of deteriorating safety culture, deferred maintenance, and management failure that was visible, partially and imperfectly, but visible, within the organisation. The Baker Panel, appointed to review BP's safety culture following the disaster, found that BP had confused personal safety performance, the metrics around slips, trips, and minor injuries, with process safety performance, the management of major hazard risk. BP had excellent personal safety metrics and poor process safety management. Its leadership believed the organisation was safe because the metrics it was measuring told them it was safe. The metrics it was not measuring, the maintenance deferrals, the near-misses in major hazard processes, the safety culture at its most hazardous sites, told a different story.

That gap between the story the metrics tell and the story the organisation's actual risk state tells is one of the defining features of pre-disaster blindness. Understanding how it develops, why it persists, and how it can be identified and closed is the subject of this book.

Towards a Different Kind of Analysis

The argument of this chapter can be stated simply: the dominant mental model of disaster causation, the blame attribution model, is inadequate to explain why major industrial disasters occur and insufficient to prevent them from recurring. It is inadequate because it focuses on the wrong level of analysis: the individual decision rather than the organisational conditions that shaped it. It is insufficient because the corrective actions it generates address symptoms rather than causes.

The alternative is not the absence of accountability. The evidence that this book examines does not suggest that individuals bear no responsibility for the decisions they make within organisations. It suggests that individual accountability, to be meaningful, must be grounded in an accurate understanding of the organisational context in which those decisions were made, including the structures, cultures, pressures, and information failures that shaped them. That understanding requires a different kind of analysis: one that examines not just what happened, but how the organisation produced the conditions that made it happen.

That is the analysis this book provides. It begins, in Chapter 2, with the concept that has done more than any other to advance our understanding of those conditions: Barry Turner's concept of the incubation period, and the question of how organisations come to be blind to the danger they are accumulating.

Chapter 2

The Incubation Problem

How disasters develop over time, why warning signals fail to register, and the enduring relevance of Barry Turner's foundational insight.

A Sociologist Looks at Disasters

In the mid-1970s, a British sociologist named Barry Turner undertook an analysis of a body of data that no one had previously thought to examine in a systematic way. Turner collected the official inquiry reports from a large sample of major British disasters and accidents, events involving mass casualties or significant harm that had occurred over the preceding two decades, and read them carefully, looking not for the proximate causes that the reports themselves identified but for the organisational and social processes that had preceded the disasters and had, in some sense, produced the conditions that made them possible.

What he found, and what he published first as a series of academic papers and then as Man-Made Disasters in 1978, was not what the inquiry reports had concluded. The inquiry reports had, in the main, produced the kind of causal analysis that inquiry processes are structured to produce: identified failures at specific points in a causal chain, attributed to specific individuals or systems, with specific recommendations for corrective action. What Turner found beneath and behind those accounts was something different: a pattern. A consistent, recurring pattern in which the conditions for disaster had been developing over an extended period, during which the relevant information, the warning signals, the known hazards, and the accumulating evidence that something was wrong, had been present within the system but had not been correctly registered, interpreted, or acted upon.

Turner called this the incubation period, the pre-disaster phase in which the conditions for catastrophe develop, unrecognised, within the organisation. He argued that disasters do not erupt from nowhere. They grow. And

DOI: 10.4324/9781042017539-4

they grow in ways that, in retrospect, are almost always visible, not clearly, not with the certainty that hindsight provides, but visible enough that a different kind of organisational attention might have detected and addressed them before they became irreversible.

Turner's framework was, at the time, a significant departure from the dominant approaches to disaster analysis, which tended to focus on technical failures, equipment malfunctions, and the immediate decisions of the individuals directly involved. His insistence on examining the extended organisational history of disasters, the period before the event rather than the event itself, was not immediately embraced by the engineering and safety communities, whose analytical frameworks were orientated toward technical causation. But over the decades that followed, as the disaster literature accumulated case after case demonstrating the same incubation pattern, Turner's foundational insight came to be recognised as one of the most important contributions to the sociology of risk and disaster (Pidgeon and O'Leary, 2000).

The Structure of Incubation

Turner's model of disaster development proposed a six-stage sequence that begins long before the precipitating event and continues through the disaster and its aftermath. Understanding this sequence is important for understanding both the strength and the limits of his framework.

The first stage is what Turner called the notionally normal starting point: the period before the incubation begins, in which the organisation is operating within its established norms and the hazard that will eventually produce disaster is either not yet present or not yet significant. This stage is important primarily as a baseline against which subsequent developments can be measured.

The second stage, and the one that contains the most analytically important material, is the incubation period itself. During incubation, a series of events occurs that, taken together, constitute the development of the disaster conditions. These events may include technical changes to systems or processes, organisational changes to structures or personnel, shifts in operational norms or cultural understandings of acceptable practice, economic pressures that alter the balance between production and safety management, and regulatory or management failures that leave known hazards unaddressed. Critically, many of these events are individually unremarkable. Each one, taken alone, might not appear to constitute a significant risk. It is their accumulation, the combination of changes to multiple systems and processes within the same organisational environment, that produces the dangerous condition (Turner and Pidgeon, 1997).

The third stage is the precipitating event: the immediate trigger of the disaster. This is the event that most inquiry processes focus on, and it is, in Turner's framework, the least analytically interesting component. The precipitating event is important because it triggers the disaster, but it is not, in most cases, the meaningful cause of it. The meaningful cause is the incubated condition that the precipitating event exposed.

The fourth, fifth, and sixth stages, the onset of the disaster, the rescue and salvage operations, and the full cultural readjustment that follows the inquiry, are concerned with the disaster's aftermath and its implications for organisational and regulatory learning. Turner was as interested in the post-disaster stage as in the pre-disaster one, because he understood that how organisations respond to disasters, what lessons they extract, how those lessons are institutionalised, and whether the underlying conditions that produced the disaster are genuinely addressed determine whether the incubation pattern will recur.

Why Warning Signals Fail

The most important question Turner's framework raises is not why disasters occur – his answer to that, the incubation model, is clear enough – but why the warning signals that are present during the incubation period fail to produce the corrective response that would interrupt the development of disaster conditions. This is the question that subsequent scholarship has most productively engaged with, and it is the question that sits at the heart of the Blindspot Model developed in this book.

Turner identified several mechanisms through which warning signals fail to produce appropriate organisational responses. The first, and perhaps the most fundamental, is what he called the rigidity of belief. Organisations, like individuals, develop belief systems about how their systems work, what constitutes acceptable operational practice, and what kinds of events are plausible within their operational environment. These belief systems are functional: they allow rapid decision-making in familiar conditions, they create shared understandings that enable coordinated action, and they provide the interpretive frameworks through which new information is assessed and responded to. But they also create a kind of cognitive inertia that makes it difficult for organisations to register information that does not fit their established understandings. Warning signals that are inconsistent with the prevailing belief system tend to be reclassified as anomalies, as measurement errors, or as irrelevant variations, rather than recognised as evidence that the belief system itself requires revision.

The second mechanism is what might be called the horizon problem. Disasters typically develop across timeframes that are longer than the attention spans of the organisational processes designed to manage risk. A hazard that develops over three years may be invisible to a system that reviews risk quarterly, because no single review captures enough of the developmental trajectory to identify the accumulating danger. This is not a failure of any individual review. It is a structural feature of risk management systems that are designed to capture cross-sectional snapshots of the organisation's risk state rather than longitudinal trajectories of risk development.

The third mechanism, and one that Turner identified as particularly important, is the problem of information in silos. During the incubation period, relevant information is typically distributed across multiple parts of the organisation. One team knows about the engineering anomaly. Another knows about the maintenance deferral. A third has recorded the near-miss. A fourth is aware of the production pressure that has been shaping operational decisions. No single individual or team has access to all of this information simultaneously. And because the organisational structures that would be needed to assemble and interpret it, cross-functional risk reviews, systematic near-miss analysis, and genuine upward communication of operational concerns are either absent or inadequate, the full picture of accumulating risk never comes together until after the disaster makes it visible.

> *The warning signals were there. They were in the incident reports, in the near-miss logs, in the conversations at shift handover, in the maintenance records, in the calculations that some engineer had done and then put aside because no one had asked for them. They were there. They were just not assembled, not interpreted, and not acted upon in time.*

The Revised Framework: Turner and Pidgeon

Turner's original 1978 work was substantially revised and updated in the 1997 second edition, produced with his colleague Nick Pidgeon following Turner's death in 1995. The revised edition incorporated two decades of subsequent disaster scholarship and addressed some of the limitations and criticisms that Turner's original framework had attracted.

One of the most important developments in the revised edition was the more explicit treatment of organisational culture as a factor in disaster incubation. The original work had focused primarily on informational and structural explanations for organisational blindness, the failure to assemble and interpret warning signals, without fully engaging with the cultural dimensions of that failure. The revised edition, drawing on the intervening literature on organisational culture and the sociology of risk, gave much greater attention to the ways in which cultural norms, shared assumptions, and institutionalised practices shape what organisations are able to see and respond to (Turner and Pidgeon, 1997).

This cultural turn was important because it pointed toward a dimension of the incubation problem that purely informational or structural explanations cannot fully capture. The question of why organisations fail to see accumulating danger is not only a question about information flows and organisational structures. It is also a question about what organisations are culturally disposed to look for, what they are culturally permitted to report, and what they are culturally able to act on. A safety culture that rewards the suppression of bad news, that treats concern about risk as a lack of commitment, or that defines professional competence in terms of the ability to manage under pressure rather than the ability to identify and address hazard, will produce blindness to warning signals regardless of the quality of its information management systems.

Pidgeon and O'Leary (2000) extended this analysis in a subsequent paper that examined the relationship between Turner's incubation model and the emerging literature on organisational culture, high-reliability organisations, and safety management systems. They argued that the cultural conditions for successful incubation, the shared assumptions, norms, and practices that allow warning signals to be systematically misinterpreted or ignored, are not exotic features of uniquely dysfunctional organisations. They are the products of ordinary organisational processes, and they develop in organisations that, by any conventional measure, are functioning well.

Incubation in Practice: Three Examples

To make the incubation concept concrete, it is worth examining briefly three cases in which the incubation pattern operated with particular clarity. Each of these cases will be examined in much greater depth in subsequent chapters. The purpose here is simply to demonstrate the generality of Turner's framework across different industries, different countries, and different eras.

Piper Alpha, North Sea, 1988

The Piper Alpha oil platform disaster killed 167 workers on the night of 6 July 1988. It remains the deadliest offshore oil and gas accident in history. Lord Cullen's subsequent public inquiry, one of the most thorough and analytically rigorous disaster inquiries ever conducted, found that the conditions for the disaster had been developing over an extended period. The permit-to-work system, which was the primary mechanism for managing the risk of simultaneous maintenance operations on the platform, had deteriorated significantly in the years preceding the disaster. The condensate pump that was the immediate trigger of the disaster had been undergoing maintenance that had not been completed at the end of the shift. The relevant information about its status, that the pump was out of service and that its pressure safety valve had been removed, was not communicated to the oncoming shift. The decision to start the pump anyway, made under production pressure in the absence of that information, triggered a gas leak that led to the explosion (Cullen, 1990).

Each of the conditions in that causal chain had an incubation history. The deterioration of the permit-to-work system had been developing for years. The production pressure that shaped the decision to start the pump had been a feature of the platform's operational culture for at least as long. The communication failures at shift handover were not unusual events. They were characteristic features of how the platform operated. The warning signals, the permit-to-work deficiencies, the near-misses, the known pressure to maximise production, were present within the system. They had not been assembled, interpreted, and acted upon in the way that would have been required to prevent the disaster.

Herald of Free Enterprise, Off Zeebrugge, 1987

On the evening of 6 March 1987, the roll-on roll-off passenger ferry Herald of Free Enterprise left the port of Zeebrugge with its bow doors open. The assistant bosun responsible for closing the doors had fallen asleep in his cabin. The officer responsible for checking that the doors were closed had left the car deck, assumed they were closed, and gone to his station. The captain, on the bridge, could not see the doors from his position and had no indicator to tell him whether they were open or closed. Within minutes of leaving port, water entered through the open doors, and the vessel capsized. One hundred and ninety-three people died.

The Sheen Inquiry found that the disaster was the product of what it memorably called a disease of sloppiness, a pervasive cultural attitude within the operating company, Townsend Thoresen, that prioritised speed and efficiency over the systematic management of safety-critical procedures. The open-door sailing was not, as it turned out, unprecedented. There had been previous incidents involving sailing with doors open or partially open, none of which had produced serious harm, and none of which had produced a systematic response from management. The possibility of sailing with open doors had been raised as a safety concern in internal correspondence before the disaster. Those concerns had not been actioned. The warning signals were present. They had not been registered as requiring urgent response (Sheen, 1987; Dekker, 2011).

Esso Longford, Victoria, Australia, 1998

The Esso Longford gas plant explosion, discussed briefly in the Introduction, provides perhaps the most clearly documented Australian example of the incubation pattern. The Royal Commission found that the conditions for the disaster had been developing for years before the explosion. The engineering knowledge required to manage the specific hazard that caused the disaster, the behaviour of heat exchangers under abnormal temperature conditions, had existed within the organisation but had not been adequately transferred to operational personnel following a restructuring that had removed resident engineers from the plant. The operational practices that contributed to the accident had deviated progressively from those originally designed for the plant. The training and competency management systems that should have identified and addressed those deviations had deteriorated. The incident reporting systems that should have flagged the accumulating near-misses had not been functioning as intended.

Hopkins' (2000) analysis of the disaster demonstrated how each of these deteriorations had developed incrementally, under the pressure of an organisational culture that emphasised cost reduction and operational efficiency, and within a management structure that had progressively reduced the technical safety oversight of the plant's operations. The disaster was not caused by any single decision or any single failure. It was the product of years of incremental change, each step individually small and collectively catastrophic.

The Limits of the Incubation Model

Turner's incubation model is foundational, but it has limits. Recognising those limits is important for understanding why this book's Blindspot Model, which builds on Turner's work, is a necessary advance rather than simply a restatement of it.

The first limitation is specificity. Turner's model identifies the general phenomenon of incubation, the pre-disaster accumulation of unrecognised warning signals, without fully specifying the organisational mechanisms through which incubation occurs. It tells us that disasters have long histories, but it does not provide a detailed account of the specific organisational conditions that produce those histories. The Blindspot Model attempts to provide that account by identifying five specific conditions, Signal Attenuation, Normalised Tolerance, Structural Secrecy, Production Gravity, and Leadership Detachment, that are the primary mechanisms through which incubation operates.

The second limitation is prescriptive insufficiency. Turner's model is primarily descriptive and analytical rather than prescriptive. It explains what happens during the incubation period and why warning signals fail to register, but it does not provide the kind of detailed practical guidance that leaders and safety professionals need to identify and interrupt the incubation process in their own organisations. The final part of this book attempts to address this limitation through the Active Foresight Framework.

The third limitation is that Turner's model was developed primarily from an analysis of British industrial disasters of the 1960s and 1970s, and his original framework was not designed to incorporate the full range of organisational, cultural, and psychological factors that subsequent scholarship has identified as relevant to pre-disaster blindness. The revised edition with Pidgeon addressed some of these limitations, but the integration of the broader disaster literature into a single, comprehensive framework was not its purpose. That integration is what this book attempts.

None of these limitations reduces the significance of Turner's contribution. His identification of the incubation phenomenon was the conceptual breakthrough that made the subsequent 50 years of disaster scholarship possible. Everything that follows in this book builds on what he established. But building on it means going further than he went, specifying the mechanisms, elaborating the framework, and providing the practical guidance that the evidence demands.

Chapter 3

Introducing the Blindspot Model

Five interlocking conditions, how they interact, and what their combination produces in organisations that cannot see the danger in front of them.

From Description to Framework

Turner gave us the phenomenon. Vaughan, Reason, Dekker, Hopkins, and the broader disaster literature gave us the mechanisms. The scholarship has done its work well, and it has done it rigorously. What remains difficult for the practitioner is not the absence of insight but the absence of integration: the task of assembling what is known, across different traditions and decades of careful scholarship, into a single, named, and practically actionable account that a safety manager, an operations director, or a board member can hold in mind while looking at their own organisation. That is the gap this book attempts to address.

The Blindspot Model is offered in that spirit. It is not advanced as a fundamentally new theoretical claim, and it does not position itself as a replacement for the scholarship it draws on. It is a structured synthesis of that scholarship, organised around five interlocking organisational conditions that, when operating in combination, reliably produce the inability to see accumulating danger in time to prevent catastrophic harm. Each condition has been identified, named, and analysed – in some cases under different terminology – by researchers working in the disaster sociology, safety science, organisational behaviour, and human factors traditions. What the model contributes is the integration of these conditions into a coherent, interacting set and the argument that it is their combination rather than the presence of any single condition that produces the most severe and most consistent forms of organisational blindness. The originality of the contribution lies in that synthesis and in its translation into practitioner-accessible form, not in the underlying theoretical raw material, which the literature has already provided with considerable rigour.

DOI: 10.4324/9781042017539-5

This chapter introduces the model in full. It defines each condition, explains its theoretical basis, identifies its characteristic indicators, and examines how it relates to the other four conditions in the model. Subsequent chapters examine each condition in depth through detailed case study analysis. The purpose here is to establish the framework clearly enough that those analyses can be read within it.

A note on scope is warranted before proceeding. The Blindspot Model is, by its diagnostic nature, a Safety-I instrument in the terms developed by Hollnagel, Wears and Braithwaite (2015): its concern is to identify the organisational conditions that precede failure so that those conditions can be interrupted. It is not, and does not attempt to be, a complete safety theory. A mature approach to active foresight – the practical orientation this book ultimately moves toward – requires that leaders hold both Safety-I and Safety-II perspectives simultaneously: asking not only 'why do we fail?' but also 'how do we succeed under varying conditions, and what makes that success fragile?' This wider framing is taken up directly in Chapter 12. The framework presented in this and the following chapters is intended to do one thing well – to make the conditions for organisational blindness visible – rather than to displace the broader safety science literature within which it sits (Figure 3.1).

Condition One: Signal Attenuation

The first condition of the Blindspot Model is Signal Attenuation: the systematic filtering, softening, and loss of safety-critical information as it moves upwards through organisational hierarchies.

The concept of Signal Attenuation draws on Turner's original account of incubation; on Hopkins' extensive analysis of how bad news travels, or fails to travel, in major hazard organisations, and on Karl Weick's work on sensemaking in organisations. Its essential claim is this: in most large organisations operating in high-hazard environments, the people who are closest to the hazard, the operators, the maintenance crews, and the frontline supervisors, have information about the state of the system that the people with the authority to make decisions about it do not have. The gap between what is known at the operational level and what is known at the decision-making level is not primarily a product of individual communication failures. It is a structural feature of hierarchical organisations operating under conditions in which bad news is implicitly or explicitly unwelcome.

Weick's (1995) analysis of sensemaking in organisations is helpful here. Weick argues that organisations do not simply receive and process information; they actively construct their understanding of their environment through ongoing processes of interpretation and meaning-making. What gets noticed, what gets reported, and what gets acted on depends not just on what is objectively happening in the system but on the interpretive frameworks that organisational members use to make sense of what they observe.

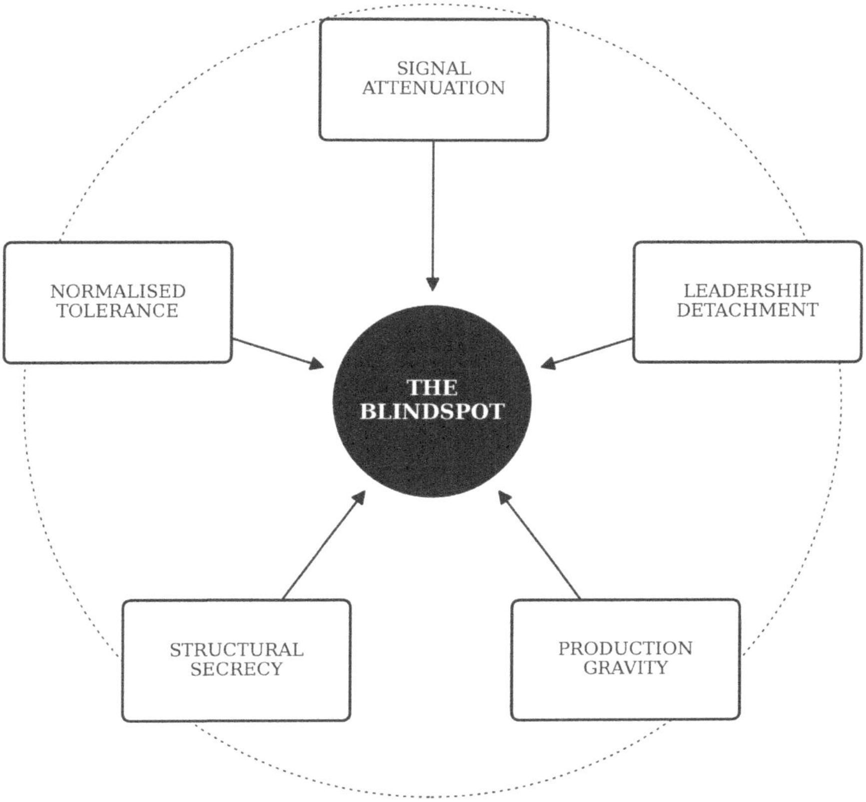

Figure 3.1 The Blindspot Model: the five organisational conditions that converge to produce the catastrophic blindspot.

When those frameworks are shaped by cultural norms that discourage the reporting of problems, by reward systems that punish bad news, by management cultures that treat concern about risk as weakness or disloyalty, by professional norms that define competence as the ability to cope without complaint, then the information about accumulating danger that exists at the operational level will be systematically filtered before it reaches the people who need to act on it.

Hopkins (2019) has documented this phenomenon across multiple disaster cases, showing how the communication structures of major hazard organisations tend to produce systematic upward filtering of safety-critical information. Managers report to senior managers. Senior managers report to executives. Executives report to boards. At each level of this reporting chain, information is aggregated, summarised, and contextualised in ways that tend to present the organisation's performance in the most favourable light consistent with the evidence. The result is not deliberate dishonesty; in most cases, the people producing those reports are acting in good faith within the cultural and institutional constraints of their roles. The result is structural inaccuracy: a leadership team whose picture of the organisation's actual risk state is systematically rosier than the reality.

The characteristic indicators of Signal Attenuation include: the consistent absence of bad news from upward reporting while problems are visible at the operational level; the normalisation of informal workarounds that are known to operators but not recorded in management systems; the tendency of near-misses and minor incidents to be resolved at the operational level without escalation; and the gap between what senior leaders believe about their organisation's safety culture and what frontline workers report when surveyed anonymously.

> *In every disaster case examined in this book, the information that would have been sufficient to interrupt the development of catastrophic conditions existed somewhere within the organisation. The problem was never a shortage of information. The problem was always a failure of the organisational mechanisms that should have moved it to the people who could act on it.*

Condition Two: Normalised Tolerance

The second condition is Normalised Tolerance: the gradual, culturally embedded process through which repeated exposure to a risk that does not immediately produce harm generates a revised, more permissive understanding of what constitutes acceptable operational practice.

This is Vaughan's (1996) normalisation of deviance, and it is one of the most consistently replicated phenomena in the disaster literature. Vaughan's original formulation, developed through her forensic analysis of the Challenger decision-making process, described how NASA's repeated experience of O-ring anomalies that did not produce catastrophic failure gradually transformed the engineers' and managers' understanding of what those anomalies meant. Each flight on which anomalies occurred but the shuttle performed successfully became, in effect, evidence that the anomalies were acceptable, that the system could tolerate them. Over time, the deviation from the design specification that the anomalies represented was reclassified, within the organisational culture, from a hazard requiring urgent attention to an acceptable operational characteristic. By the time of the Challenger launch decision, the O-ring anomalies that should have been understood as evidence of a potentially catastrophic design vulnerability were being treated by the people responsible for the launch decision as a known and manageable feature of shuttle operations.

Vaughan was careful to distinguish normalisation of deviance from deliberate risk-taking. The NASA engineers and managers who participated in the Challenger launch decision were not indifferent to the safety of the crew. They were people who had come, through a long process of organisational experience and cultural learning, to understand their situation in a particular way, a way that made the launch decision appear not reckless but reasonable, not dangerous but within the bounds of acceptable uncertainty. This is what makes normalisation of deviance so difficult to detect and so difficult to interrupt from inside the organisation. It does not feel like the acceptance of unacceptable risk. It feels like experienced, informed professional judgement.

The normalisation process typically operates through a specific mechanism that can be described as the near-miss reclassification cycle. An event occurs that is, objectively, a warning signal, a near-miss, an anomaly, or a minor incident involving a process or system that carries major hazard potential. The event does not produce serious harm. It is investigated, or recorded, or discussed. A conclusion is reached that is consistent with the prevailing belief system: the anomaly was within acceptable limits, the near-miss was a one-off, and the minor incident was addressed by the immediate corrective action taken. The boundary of acceptable practice is imperceptibly extended. The next similar event is assessed against this revised boundary rather than the original one. Over time, what began as a clear deviation from safe operating conditions has become the new normal.

Dekker (2011) describes this as the gradual erosion of safety margins through successive locally rational decisions, each of which appears justified in the immediate context but which, in combination, produces an organisation operating dangerously close to its limits. He argues that this process is not a malfunction; it is the ordinary product of organisations optimising for performance under conditions of competitive pressure and incomplete information. That is what makes it so difficult to see from inside. It looks like the system is working as intended.

Condition Three: Structural Secrecy

The third condition is Structural Secrecy: the fragmentation of safety-critical information across organisational boundaries, functional silos, geographic sites, and hierarchical layers, such that no individual or team has access to the complete picture of accumulating risk.

Vaughan (1996) coined the term Structural Secrecy to describe the way in which information in large, complex organisations is distributed across so many boundaries and filtered through so many interpretive layers that the complete picture of any given risk situation is effectively unavailable to any single decision-maker. The secrecy is structural rather than intentional: it is not the product of deliberate concealment but of the ordinary fragmentation of information in complex systems.

In the context of major hazard risk management, Structural Secrecy operates with particular force because the information relevant to a developing hazard is typically held by different groups in different parts of the organisation. The engineering team knows about the design limitation. The operations team knows about the procedural workaround that has developed in response to that limitation. The maintenance team knows about the deferred repairs that have been accumulating on the relevant equipment. The safety team knows about the near-miss reports that have been filed in the preceding months. The finance team knows about the budget constraints that have been shaping the maintenance programme. No single person or team has all of this information. And unless there is a specific, functioning mechanism for assembling it, a cross-functional risk review, and a major hazard register that is actively maintained and regularly reviewed by people with the authority to act on it, the complete picture of the developing hazard never comes together.

Weick et al. (1999) have argued that the capacity to assemble fragmented information into a coherent picture of the system's state is a defining characteristic of high-reliability organisations. High-reliability organisations, the organisations that operate complex, high-hazard systems with consistently lower rates of catastrophic failure than their counterparts, invest heavily in what Weick and his colleagues call collective mindfulness: the organisational practices and cultural norms that enable distributed information to be aggregated, challenged, and acted on at the level of the system rather than the level of the individual component. The absence of this capacity is not a neutral condition. It is the active presence of Structural Secrecy, and Structural Secrecy is a condition in which the full picture of accumulating risk is effectively invisible to the organisation even when all of its components are present within it.

The characteristic indicators of Structural Secrecy include: functional silos that prevent cross-disciplinary communication about risk; geographic or organisational separation between technical experts and the operational environments they are responsible for; the fragmentation of safety-critical information across multiple systems, databases, or reporting processes that are not integrated; and the consistent finding, in post-disaster investigations, that the complete picture of the developing hazard was available within the organisation but was held in parts by different teams who did not communicate with each other.

Condition Four: Production Gravity

The fourth condition is Production Gravity: the constant economic and operational pressure that pulls organisational decision-making toward the performance of productive work and away from the maintenance of safety margins, producing incremental, individually justified erosions of those margins that, in combination, leave the organisation without adequate protection against catastrophic failure.

The concept of Production Gravity draws on Dekker's (2011) account of drift into failure, on Hopkins' repeated observations about the role of production pressure in major hazard disasters, and on the broader organisational behaviour literature on the relationship between performance pressure and risk-taking. Its essential claim is that the economic and operational incentives operating within any productive organisation create a systematic bias toward decisions that serve production goals at the expense of safety margins and that this bias produces, over time, exactly the kind of incremental erosion of protection that Turner's incubation model describes.

The gravitational metaphor is deliberately chosen. Gravity does not require a specific actor or a specific decision. It operates continuously on every object in its field, pulling in a constant direction. Production Gravity operates the same way. It does not require managers to make explicitly unsafe decisions. It operates through the accumulation of individually reasonable decisions: to defer a maintenance item that does not present an immediate hazard, to accept a slightly higher operating parameter that is within the technical specification, and to reduce the resources allocated to a safety function whose value is difficult to quantify, each of which makes sense in the immediate economic and operational context but which together move the organisation progressively closer to the boundary beyond which catastrophic failure becomes possible.

Hopkins (2008) documented this process with particular clarity in his analysis of the BP Texas City disaster, showing how the culture of cost reduction that BP had systematically instilled across its refinery operations had produced, at Texas City, a site in which maintenance had been systematically deferred, safety-critical equipment had been allowed to deteriorate, and the process safety management systems that should have identified and arrested these deteriorations had been stripped of the resources and authority required to function effectively. No single individual at Texas City decided to make the refinery unsafe. Many individuals, at many levels of the organisation, made many individual decisions that each appeared justifiable within the constraints of their role and which collectively produced a site that was, by any objective assessment, dangerously unsafe.

Production Gravity is particularly difficult to address because its operation is culturally invisible. In organisations where production pressure is strong and consistent, the bias toward production-favouring decisions becomes the background condition of organisational life, the way things work, the obvious thing to do, and the professional norm. Challenging it requires individuals to act against the grain of the organisation's dominant culture, to question decisions that are being made in ways that everyone around them appears to regard as normal, and to accept the professional and reputational costs that such challenges typically impose. These are not trivial requirements, and the evidence from the disaster literature suggests they are requirements that most organisations systematically fail to support.

Condition Five: Leadership Detachment

The fifth condition is Leadership Detachment: the structural loss of contact between senior leaders and operational reality that is a predictable consequence of organisational scale, hierarchy, and information filtering, producing leadership teams whose confidence in their organisation's safety performance is grounded not in accurate knowledge of the system's actual state but in the picture that the system's reporting structures have constructed for them.

Hopkins (2007) introduced the concept of mindful leadership in his analysis of the Gretley mine disaster, arguing that the leaders who are most effective in managing major hazard risk are those who actively and persistently probe beneath the surface of the formal reporting systems, who seek out the operational reality of their organisations rather than relying on the sanitised versions that those organisations' communication structures tend to produce, and who maintain genuine intellectual engagement with the gap between systems as designed and systems as actually operating. Mindful leaders, in Hopkins' account, are not distinguished primarily by their technical knowledge or their personal commitment to safety; they are distinguished by their epistemic habits: the way they seek, process, and act on information about their organisations' actual risk states.

Leadership Detachment is the opposite of mindful leadership, and it is the natural tendency in large organisations. As organisations grow in size and complexity, the structural distance between senior leaders and the operational environments they are responsible for increases. The reporting systems that carry information upward through the hierarchy become longer and more complex, creating more opportunities for filtering, summarisation, and the subtle transformation of information that moves it away from operational reality and towards institutional reassurance. The metrics that leaders use to assess their organisations' performance, the lag indicators of safety performance that dominate most executive reporting systems, capture what has already happened rather than what is developing. They tell leaders about yesterday's incidents rather than tomorrow's catastrophe.

Weick and Sutcliffe (2001) argue that high-reliability organisations are distinguished, in part, by their leaders' commitment to what they call sensitivity to operations: a genuine, active, and continuous awareness of the gap between the organisation's formal systems and its actual operational practice. This sensitivity is not naturally produced by normal organisational processes. It requires specific structures, regular visits to operational sites, direct conversations with frontline workers, systematic processes for identifying and examining the gap between policy and practice, and specific cultural norms that make it safe for operational personnel to communicate honestly about what is actually happening in the field.

The characteristic indicators of Leadership Detachment include reliance on aggregate lag indicators as the primary means of assessing safety performance; the absence of systematic mechanisms for direct communication between senior leaders and operational personnel; the consistent surprise of senior leaders when serious incidents are revealed to have been preceded by extended periods of deteriorating conditions; and the gap between what senior leaders report believing about their organisation's safety culture and what operational personnel report experiencing.

How the Five Conditions Interact

The Blindspot Model is not a simple list of five independent risk factors. The conditions interact with each other in ways that are important to understand, because it is the interactions, the ways in which each condition reinforces and enables the others, that produce the most dangerous and most persistent forms of organisational blindness.

Signal Attenuation and Leadership Detachment reinforce each other directly. An organisation in which safety-critical information is filtered before it reaches decision-makers will produce leaders who are detached from operational reality, not because those leaders are disengaged, but because the information available to them does not accurately represent the system's actual state. And an organisation whose leaders have become detached from operational reality will tend, over time, to develop the cultural and structural conditions that produce Signal Attenuation, because leaders who are not genuinely engaged with operational reality do not create the demand for honest upward communication that would interrupt the filtering process.

Normalised Tolerance and Production Gravity interact through the near-miss reclassification cycle described earlier. Production pressure creates the incentive to continue operating in conditions that a more conservative safety culture would halt. Each time the organisation continues operating in those conditions without a catastrophic outcome, the normalisation process extends the boundary of what is considered acceptable. The extended boundary then reduces the resistance to the next increment of production-gravity-induced risk-taking. Over time, the two conditions together produce a progressive and self-reinforcing erosion of safety margins that is invisible from inside the organisation because every individual step in the process appeared, at the time it was taken, to be within the revised understanding of what was acceptable.

Structural Secrecy amplifies all four of the other conditions. When information about accumulating risk is fragmented across organisational boundaries, the effects of Signal Attenuation are multiplied, not only does the information fail to travel upward, it also fails to travel laterally to the people who might be able to assemble the complete picture. When Structural Secrecy prevents the full extent of Normalised Tolerance from being visible to anyone within the organisation, the correction mechanisms that might interrupt the normalisation process, a risk review that assembles the complete picture of procedural deviations, and a safety audit that identifies the full extent of maintenance deferrals cannot function effectively. And when Structural Secrecy prevents senior leaders from receiving the information they would need to detect their own detachment, it deepens and extends the Leadership Detachment that the Signal Attenuation condition has already produced.

The result of all five conditions operating in combination is an organisation that is, in a very precise sense, blind to its own risk state. It is not an organisation that is deliberately taking unacceptable risks. It is an organisation that has lost the structural, cultural, and informational capacity to see, accurately and completely, what kind of risks it is actually taking. And that blindness, systematic, organisationally produced, and self-reinforcing, is what this book calls the Blindspot.

The Blindspot is not a gap in the organisation's knowledge. It is a gap in the organisation's capacity to see. The information exists. The systems exist. The people exist. What is missing is the organisational capacity to assemble, interpret, and act on what is already there. That capacity, active foresight, is what this book's final section addresses.

The Model and the Evidence

A framework that cannot be tested against evidence is not a framework; it is a speculation. The Blindspot Model is grounded in the existing literature and is offered as a synthesis of that literature, not as an invention of the author. But its value as a framework depends on its ability to explain the evidence from actual disasters more accurately and more usefully than alternative accounts.

The following five chapters test this claim. Each chapter takes one condition of the Blindspot Model and examines it in depth through detailed analysis of two or more major disasters in which that condition can be clearly identified. The analyses draw on official inquiry reports, peer-reviewed academic analyses, and the best available narrative accounts of each disaster. They are intended to demonstrate not simply that the condition in question was present in the case but how it developed, how it interacted with the other conditions in the model, and what its characteristic indicators were in each specific organisational context.

The reader who works through those analyses will, I hope, emerge with a clearer sense not just of why those specific disasters happened but also of how to recognise the conditions that produce disasters in their own organisations in real time before the precipitating event that transforms incubated conditions into irreversible harm.

That, in the end, is the purpose of the framework. Not to explain what happened. To help prevent what is developing.

Part II

The Five Conditions

Chapters 4–8

The conditions do not operate in isolation. They operate in organisations. That is what makes them so difficult to see.

Part II examines each of the five conditions of the Blindspot Model in depth. For each condition, the analysis proceeds in the same way: theoretical grounding in the disaster scholarship; demonstration of the condition's operation through detailed analysis of two major cases; identification of the characteristic indicators through which the condition can be detected; and practical guidance on how it can be addressed.

The cases examined in Part II span five decades, four countries, and five distinct industries. The patterns they reveal are consistent. The five conditions are not artefacts of particular industries or regulatory environments. They are features of how large organisations operating under production pressure manage safety-critical information and decision-making. Understanding how each condition works is the necessary foundation for the integrated analysis of Part III and the practical framework of Part IV.

DOI: 10.4324/9781042017539-6

Chapter 4

Signal Attenuation

When the Warning Never Arrives

How safety-critical information is filtered, softened, and lost before it reaches the people responsible for acting on it, Deepwater Horizon and Grenfell Tower.

The Communication of Danger

In every organisation examined in the disaster literature, the information that would have been sufficient to prevent catastrophe was, at some point before the disaster, present within the system. It was held by operators who had noticed something wrong. It was contained in incident reports that had been filed and not escalated. It was embedded in the professional concern of engineers who had run calculations that worried them. It was visible in the maintenance records of people who knew that equipment was not performing as it should. The information was there. What was absent was the organisational capacity to move it, accurately, completely, and in time, to the people with the authority and the responsibility to act on it.

Signal Attenuation is the name this book gives to the organisational process by which that movement fails. It is not a single event. It is not, in most cases, a deliberate act of suppression. It is a structural tendency built into the hierarchies, cultures, incentive systems, and communication practices of large organisations to filter, soften, reframe, and ultimately lose the safety-critical information that originates at the operational level before it reaches the decision-making level where corrective action is possible.

The word attenuation is drawn from physics, where it describes the reduction in the strength of a signal as it travels through a medium. An attenuated signal is not absent; it is weakened. In organisational terms, an attenuated safety signal is one that has been weakened sufficiently by its passage through the organisational medium, through layers of reporting, cultural pressures against bad news, and structural incentives for reassurance, that it no longer carries the force required to produce corrective action. The signal arrives.

DOI: 10.4324/9781042017539-7

But it arrives too weak to be heard above the background noise of normal organisational operations.

This chapter examines Signal Attenuation in depth through two detailed case studies: the Deepwater Horizon blowout of 2010 and the Grenfell Tower fire of 2017. These cases are chosen because they demonstrate the operation of Signal Attenuation across two very different organisational contexts: one, a large multinational corporation operating a high-technology offshore drilling programme, the other, a public housing management system operating in an urban residential context. The consistency of the pattern across these different contexts is itself a significant finding, and it is one that carries direct implications for the way the condition is understood and addressed.

Sense-Making on the Ground

The academic literature describes this process using the language of sense-making (Weick, 1995): the continuous, collective work through which people in organisations construct a shared account of what the facts in front of them actually mean. The phrase is easy to pass over as a theoretical term. It is worth stopping on, because sense-making is not an academic abstraction. It is what a frontline supervisor is doing every time she looks at a slightly unusual gauge reading and decides whether it is a problem, a quirk, or a sign that yesterday's repair did not hold. It is what a safety professional is doing when he reads the fourth near-miss report this month from the same workface and has to decide whether to write it up as a developing trend or file it as an isolated event. It is what a mid-level manager is doing when her engineer comes to her with a concern about a piece of equipment that the schedule cannot afford to have shut down.

In each of those cases, the individual is not operating on facts alone. They are operating on facts filtered through a framework for deciding what those facts mean – a framework they have absorbed from the conversations, priorities, pressures, and rewards of the organisation they work in. If that framework has been shaped by the cultural pressures this chapter describes – the systematic preference for reassurance, the normalisation of informal workarounds, and the tacit understanding that bad news slows things down – then the sense-making process will tend, without anyone deciding that it should, towards the interpretation that permits operations to continue. The gauge reading is explained. The near-miss pattern is reframed. The engineer's concern is recorded but not escalated. None of this requires bad faith. It requires only that the organisation has taught its people, over time and by consistent signals, how to see.

Sense-making, in other words, is the thing that fails first. By the time an information flow has been formally attenuated in a report or a briefing, it has usually already been attenuated in the mind of the person who decided what to put in the report. The practical importance of the concept lies in that earlier moment – and in the recognition that interrupting Signal Attenuation at the documentation stage will never be sufficient if the sense-making habits of the people generating those documents have already been trained in the wrong direction.

Case Study One: Deepwater Horizon, Gulf of Mexico, 2010

Case Overview: *On 20 April 2010, the Deepwater Horizon drilling platform experienced a blowout of the Macondo well, resulting in an explosion that killed 11 workers and triggered the largest accidental marine oil spill in history. The disaster was preceded by an extended period in which safety-critical information about the well's condition, the integrity of its barriers, and the adequacy of its well control procedures was present within the system at multiple levels but failed to reach the decision-makers in a form that produced appropriate corrective action.*

The Organisational Context

The Macondo well was a challenging well by any measure. During the drilling process it had experienced a series of complications: lost returns, stuck pipe, and an unexpected influx of gas that were individually managed but that, in combination, had produced a well significantly more complex and hazardous than had been anticipated at the planning stage. By the time it reached the final critical phase, the temporary abandonment procedures being executed on 20 April, it had been drilled over schedule and over budget. Commercial pressure to complete the abandonment and release the Deepwater Horizon rig for its next assignment was substantial and explicitly acknowledged in the internal communications of the organisations involved (National Commission on the BP Deepwater Horizon Oil Spill and Offshore Drilling, 2011).

The well abandonment procedure required the installation of a series of barriers designed to prevent the uncontrolled flow of hydrocarbons from the reservoir. The adequacy of those barriers, the number required, their placement, and the tests used to confirm their integrity were the subject of a series of decisions made in the days and hours before the disaster. Each decision involved a trade-off between the safety imperative of maintaining robust barrier protection and the commercial imperative of completing the abandonment efficiently. In each case, the decision favoured efficiency. And in each case, the information that might have supported a more cautious decision was present somewhere in the system but was not assembled, interpreted, or acted on in the way that the circumstances required.

The Negative Pressure Test and Its Failure

The most critical safety signal in the Macondo sequence was the result of the negative pressure test conducted on the afternoon of 20 April. A negative pressure test is designed to confirm that the barriers installed in the well are sufficient to prevent flow from the reservoir. The test conducted on the Macondo well produced anomalous results, specifically, a pressure reading on the kill line that was inconsistent with a successful test. This anomalous reading was, objectively, a clear warning signal: it indicated that the well's barriers were not providing the protection required before the drill string could be removed and the well temporarily abandoned.

The anomalous reading was observed by the Transocean well site leaders and the BP company men on the platform. It was discussed. A series of explanations was proposed and evaluated. The explanation ultimately accepted, that the anomalous reading was caused by a phenomenon known as the bladder effect, in which fluid trapped in the drill string was influencing the pressure reading on the kill line, had no technical basis and had never been proposed in any documented well control training or guidance as an explanation for this type of pressure anomaly (Bly, 2010). It was, in Hopkins' (2012) analysis, a post-hoc rationalisation: an explanation constructed to make an anomalous result consistent with the belief that the test had been passed, rather than a genuine technical evaluation of what the result actually meant.

The signal, the anomalous pressure reading, was attenuated in real time, in a room where the people present had the information, the authority, and the nominal responsibility to act on it. What they lacked was the organisational and cultural context in which acting on it would have been the normal, expected, professionally supported response. The commercial pressure was immediate and tangible. The safety concern was abstract and uncertainly framed. The cultural norm, in an environment where schedule pressure was intense and where professional identity was closely bound up with the ability to manage complexity and maintain operational momentum, favoured the interpretation that allowed work to continue.

Hopkins (2012) identifies this as a specific and recurring form of Signal Attenuation: the in-context reclassification of warning signals from indicators of unacceptable hazard to manageable anomalies, driven by the combination of production pressure, cultural norms against delay, and the ambiguity inherent in interpreting complex technical data in real time. The people in the room on the afternoon of 20 April were not incompetent. They were not indifferent to the safety of the crew. They were operating within an organisational context that had, over time, produced exactly the interpretive habits and cultural norms that made their decision to accept a technically unfounded explanation and proceed the path of least resistance.

The Deeper History of Attenuation

The Signal Attenuation that occurred on 20 April was not an isolated event. It was the final manifestation of a pattern of attenuated communication that had been operating throughout the Macondo well project and, more broadly, within BP's operational culture in the Gulf of Mexico.

The Presidential Commission's report documented multiple instances in the months and weeks preceding 20 April in which safety-critical concerns about the Macondo well, about the integrity of its casing design, the adequacy of its centraliser placement, and the number and positioning of its barriers, had been raised by engineers within the system and had not produced the responses those concerns warranted. In some cases, concerns were raised and effectively overridden by decisions made at higher levels of the organisation without adequate engagement with their technical basis. In others, concerns were framed in ways that softened their urgency, transformed from clear recommendations against a proposed course of action into qualified professional opinions that left the decision-maker room to proceed (National Commission, 2011).

One of the most significant documented examples involved the number of centralisers, devices used to ensure the casing string runs centrally in the wellbore, which is critical to the quality of the cement job that forms one of the primary well barriers. BP's drilling engineer had recommended using 21 centralisers. Only six were available on the rig.

An engineering analysis commissioned to assess the risk of proceeding with six centralisers concluded that this approach had a high probability of producing channelling in the cement, a condition that would compromise the cement barrier's integrity. That analysis was received by the relevant BP manager, who acknowledged it but decided to proceed with six centralisers on the grounds that the cement job would be followed by a cement evaluation log that would identify any channelling. The cement evaluation log was subsequently not run. The signal, the engineering analysis identifying a high probability of a compromised cement barrier, had been received, noted, and effectively discounted (National Commission, 2011).

The National Commission found that BP's organisational culture, as it operated in the context of deepwater drilling operations, did not provide the structural or cultural support for the kind of assertive upward communication of safety concerns that the management of major hazard risk requires. The incentive systems, financial, professional, and reputational, favoured the completion of wells on schedule and within budget. The communication structures did not create robust pathways for operational-level safety concerns to reach senior decision-makers in a form that carried sufficient authority to interrupt commercial momentum. The result was an organisation in which warning signals were repeatedly present and repeatedly attenuated before they could produce the corrective response they warranted.

Leadership and the Limits of Visibility

One of the most important findings of the various Macondo investigations concerned the information available to BP's senior leadership about the state of the well and the decisions being made about it. The evidence suggests that senior BP executives had limited knowledge of the specific well design and abandonment decisions that contributed to the disaster. This was not because that information was being concealed from them. It was because the organisational structures and communication practices of BP's deepwater drilling operations did not routinely transmit that level of operational detail to the senior leadership level.

This finding illustrates the relationship between Signal Attenuation and the fifth condition of the Blindspot Model, Leadership Detachment, with particular clarity. When the communication structures of an organisation systematically filter operational-level safety information before it reaches senior leaders, those leaders cannot make informed judgements about the risk states of the operations they are nominally responsible for. They are not detached because they are disengaged. They are detached because the information available to them has been attenuated to the point where it no longer accurately represents the operational reality from which it originated. The two conditions, Signal Attenuation and Leadership Detachment, are, in this sense, mutually reinforcing and mutually producing.

Case Study Two: Grenfell Tower, London, 2017

Case Overview: *On 14 June 2017, a fire that began in a fourth-floor kitchen flat of Grenfell Tower, a 24-storey residential block in west London, spread rapidly up the exterior of the building via recently installed cladding and killed 72 residents. The disaster was preceded by years during which safety concerns about the cladding, the fire safety systems, and the management practices of the building's management organisation had been raised by residents and had failed to produce adequate responses from the responsible authorities.*

The Resident Voice and Its Systematic Failure

The Grenfell Tower disaster is, in important respects, a case study in a specific and particularly troubling form of Signal Attenuation: the systematic failure of a public organisation to receive, register, and act on safety signals transmitted by the people most directly at risk, the building's own residents.

The Grenfell Action Group, a residents' organisation established to advocate for the interests of tower residents, had been documenting and publicising safety concerns about the building and its management for several years before the fire. In November 2016, seven months before the fire, the Group published a blog post that described with alarming specificity their concern that a serious fire in the tower would be catastrophic. It identified particular maintenance and management failures that they believed increased that risk, and it stated directly that they feared it would take a serious incident before the relevant authorities would respond to their concerns. The post was prescient in ways that are, in retrospect, deeply distressing to read.

The concerns the Group had been raising, about the adequacy of fire safety systems, about the quality of recent refurbishment work, and about the management practices of the Kensington and Chelsea Tenant Management Organisation (KCTMO), had been communicated through every formal channel available to them. They had written letters. They had attended management meetings. They had filed formal complaints. They had engaged with local councillors and the council itself. The signals had been transmitted, repeatedly and clearly, across an extended period. They had been attenuated, filtered, acknowledged without action, or actively dismissed, at every point in the system that should have received and acted upon them (Grenfell Tower Inquiry, Phase 1 Report, 2019).

The Grenfell Tower Inquiry found that the KCTMO's responses to resident safety concerns were, across the period examined, characterised by an institutional disposition to manage and contain those concerns rather than to genuinely investigate and act on their substance. Residents who raised safety issues were provided with responses that were, formally, responses; acknowledgements were sent, meetings were held, and assurances were provided, but they were not grounded in genuine investigation of the concerns raised. The signal was received at the organisational boundary. It was processed in ways that satisfied the formal requirements of a response. And it did not produce the substantive corrective action that the substance of the concerns warranted.

The Cladding Decision and Its Suppressed Warnings

The refurbishment of Grenfell Tower, completed in 2016, involved the installation of an external cladding system composed of materials that subsequent investigation established as combustible, materials that, when ignited, produced the rapid external fire spread that transformed a kitchen fire into a disaster of the scale that occurred. The specific panels selected, aluminium composite panels with a polyethylene core, had been the subject of documented fire safety concerns in the construction and building regulation professions for some years before the Grenfell refurbishment. Their behaviour in fire, and the limitations of the testing and certification processes used to assess that behaviour, had been the subject of professional discussion and regulatory consideration.

None of these concerns reached the decision-makers responsible for the Grenfell refurbishment in a form that interrupted the specification and installation of the cladding system. The procurement process that selected the particular panels was driven significantly by cost considerations; the polyethylene-core panels were cheaper than the fire-resistant alternative that had also been considered. The regulatory and certification framework that should have provided reliable assurance about the fire performance of the selected system was, as the Inquiry established, inadequate for the purpose to which it was being applied, and there is evidence in the documentary record that those responsible for the procurement were aware, or should have been aware, of its limitations (Grenfell Tower Inquiry, Phase 2 Report, 2024).

The Phase 2 Report found that the cladding and insulation products used on Grenfell Tower were dangerous and should not have been used. It also found that the manufacturers of those products had engaged in deliberate strategies to misrepresent their fire performance to the market and to the regulatory system, strategies that constituted, in the Inquiry Chair's assessment, a form of systematic dishonesty that contributed directly to the conditions that produced the disaster. This finding adds a dimension to the Grenfell Signal Attenuation story that is not present in most of the industrial cases examined in this book: the deliberate manufacturing and transmission of false safety signals, which crowded out the accurate signals that might otherwise have interrupted the procurement decision.

The Night of the Fire: Attenuation in Real Time

The Signal Attenuation that had operated in the years before the Grenfell fire continued to operate on the night of the fire itself, with catastrophic consequences for the residents who had been advised to stay in their flats.

The London Fire Brigade's operational response to the Grenfell fire was governed by a stay-put policy: the standard advice for residents of purpose-built residential tower blocks, based on the principle that such buildings are designed to contain a fire within the flat of origin through compartmentalisation and that residents are safer remaining in their flats than evacuating through smoke-filled common areas. This policy was appropriate for a building performing as designed. It was catastrophically inappropriate for a building whose external cladding system was acting as a continuous fire path, destroying the compartmentalisation that the stay-put strategy depends on within minutes of the fire's external spread.

The signal that the building was not performing as designed, that the fire was spreading externally in a way that made the stay-put advice dangerous rather than protective, was visible from the earliest minutes of the external fire spread. It was visible to residents watching from their windows and to firefighters on the ground. The communication systems that should have allowed that signal to interrupt the stay-put advice, and trigger the evacuation that might have saved lives, were inadequate to the task. The Inquiry found that the decision to issue an evacuation instruction was delayed significantly beyond the point at which the evidence available to the incident commander should have made the inadequacy of the stay-put strategy apparent (Grenfell Tower Inquiry, Phase 1 Report, 2019).

This real-time attenuation of a life-critical signal, the failure of the operational communication systems to translate the visible evidence of the building's catastrophic performance into a timely decision to evacuate, is a reminder that Signal Attenuation is not only a phenomenon of the months and years before a disaster. It operates during disasters as well, and its consequences in those contexts are immediate and irreversible.

The Anatomy of Signal Attenuation: Four Mechanisms

Drawing on the Deepwater Horizon and Grenfell cases, and on the broader disaster literature, four characteristic mechanisms through which Signal Attenuation operates can be identified. These mechanisms are not mutually exclusive; they typically operate in combination, but distinguishing them is analytically useful because each requires a different kind of organisational response.

Mechanism One: Cultural Filtering

Cultural filtering is the implicit or explicit discouragement of bad news within organisational hierarchies. In organisations where the professional norm is that good managers manage, where the reporting of problems is read as an indication of inadequate performance rather than appropriate safety awareness, the individuals who observe warning signals face a genuine professional incentive to manage those signals at the level at which they are observed rather than escalating them. This is not primarily a failure of individual courage. It is a rational response to a cultural environment that penalises the bearer of bad news and rewards the appearance of control.

Cultural filtering is particularly difficult to address because it operates through informal norms rather than formal policies. An organisation's formal communications policy may explicitly encourage the reporting of safety concerns. Its actual cultural norms may systematically discourage it. The gap between formal policy and cultural reality is itself a form of Signal Attenuation, one that is invisible to leaders who assess their organisation's safety communication practices by reference to the formal systems rather than the cultural experience of the people operating within them.

Mechanism Two: Interpretive Reclassification

Interpretive reclassification is the in-context transformation of warning signals from indicators of unacceptable hazard to acceptable anomalies, typically driven by a combination of production pressure and the ambiguity inherent in interpreting complex operational data in real time. This is what happened with the negative pressure test on the Deepwater Horizon. The signal was received. It was subjected to an interpretive process. It was reclassified as something other than what it objectively was, and it arrived at the decision-making level in its reclassified form, as a resolved anomaly rather than an unresolved warning.

Interpretive reclassification is facilitated by genuine complexity and genuine uncertainty. In high-technology operational environments, the data produced by monitoring systems is rarely unambiguous. Multiple interpretations of anomalous readings are, in many cases, technically possible. The question is not whether alternative interpretations exist – they almost always do – but what the cultural and institutional norms of the organisation are that determine which interpretation is adopted in conditions of genuine uncertainty. In organisations where the dominant norm favours production continuity, the interpretation adopted will tend to be the one that allows work to continue. That tendency is Signal Attenuation by a different name.

Mechanism Three: Structural Distance

Structural distance is the physical, organisational, and hierarchical separation between the people who observe warning signals and the people who have the authority to act on them. As organisations grow in size and complexity, as their operations become geographically dispersed, and as their hierarchies become more elaborate, the number of interpretive steps through which a signal must pass before it reaches a decision-maker increases, and with each step, the probability of attenuation increases as well.

Structural distance is a particularly important mechanism in the context of major hazard industries, where the people with the most direct knowledge of operational risk, the workers in the field, the operators on the plant, and the crew on the platform are often separated by several layers of management hierarchy from the people with the authority to make the decisions that would address that risk. Bridging that structural distance requires deliberate organisational investment: in communication structures that create direct pathways between operational knowledge and senior decision-making authority, in leadership practices that actively seek out operational reality rather than waiting for it to arrive through normal reporting channels, and in cultural norms that make the escalation of safety concerns an act of professional competence rather than an admission of failure.

Mechanism Four: Institutional Self-protection

Institutional self-protection is the tendency of organisations facing external scrutiny, regulatory, political, or reputational, to manage the outward presentation of their risk status in ways that prioritise institutional reassurance over accurate communication of safety concerns. This mechanism was particularly visible in the Grenfell case, where the KCTMO's responses to resident concerns appear, across the evidence examined by the Inquiry, to have been shaped as much by the management of institutional relationships and reputational exposure as by genuine engagement with the safety substance of those concerns.

Institutional self-protection operates at the boundary between the organisation and its external environment, regulators, insurers, investors, and the public, but its effects are felt internally as well. When the organisation's external communication about its safety performance is more reassuring than its internal reality warrants, it creates pressure on internal communication to align with the external story. The result is a progressive compression of the space within the organisation in which honest assessment of the actual risk state is possible.

> *In both cases examined in this chapter, Deepwater Horizon and Grenfell, the information required to prevent the disaster was present within the system. In both cases, the organisational mechanisms that should have moved that information to the people who could act on it had been degraded by cultural pressure, by interpretive norms, by structural distance, and by institutional self-protection to the point where the signal that arrived at the decision-making level bore only a passing resemblance to the signal that had originated at the operational level. That degradation is Signal Attenuation. And it is preventable.*

Identifying Signal Attenuation in Your Organisation

The practical value of the Signal Attenuation framework depends on the ability of leaders and safety professionals to identify its presence in their own organisations before it produces catastrophic consequences. The following indicators are drawn from the case analyses and from the broader disaster literature. They are not a definitive checklist; Signal Attenuation manifests differently in different organisational contexts, but they provide a starting point for the kind of honest organisational assessment that active foresight requires.

The first indicator is the consistent absence of safety concerns from formal upward reporting in organisations where operational-level personnel, when asked directly, report genuine concerns. The gap between what the formal reporting system says and what the workers say when asked directly is a direct measure of the degree to which cultural filtering is operating.

The second indicator is the pattern by which near-misses and minor incidents are resolved at the operational level without escalation. If the organisation's incident reporting data shows a high ratio of minor incidents resolved at the operational level relative to the number escalated to management review, this may indicate that the cultural norms around escalation are producing attenuation at the first level of the hierarchy.

The third indicator is the response of the organisation to safety concerns raised by people outside the management hierarchy, workers, residents, contractors, and customers. In organisations where Signal Attenuation is operating through institutional self-protection, concerns from these sources will tend to be managed rather than investigated: acknowledged formally, processed through complaint management systems, and resolved in ways that satisfy the formal requirement for a response without genuinely engaging with the substance of the concern.

The fourth indicator is the quality of the information available to senior leaders about the operational risk state of the organisation. Leaders who cannot, when asked, describe with specificity and accuracy the current major hazard risk profile of their operations, who rely on aggregate lag indicators and formal assurance reports rather than direct knowledge of the system's actual state, are exhibiting the symptoms of Signal Attenuation at the leadership level. That is not necessarily evidence of their individual failure. It is evidence that the information systems that should be providing them with an accurate picture of operational reality are not functioning as required.

These indicators are not evidence of disaster in themselves. They are evidence that the conditions for Signal Attenuation are present, that the organisational mechanisms through which warning signals travel are not functioning with the reliability that the management of major hazard risk demands. That is what makes them worth identifying and addressing. The cost of addressing

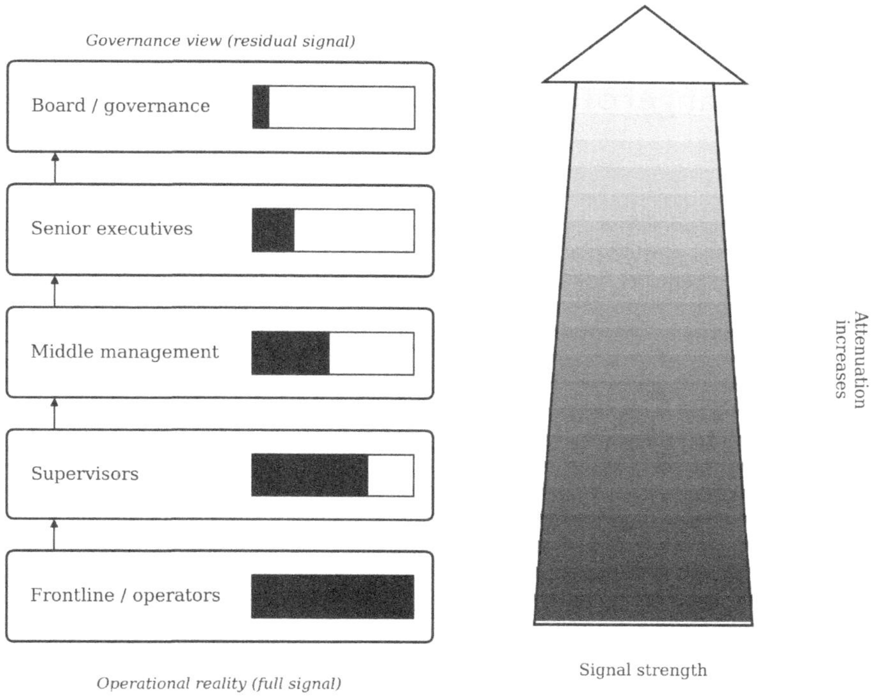

Figure 4.1 Signal Attenuation through organisational layers: how the strength of a safety-critical signal decays as it moves from the frontline to governance.

Signal Attenuation is the cost of improving communication systems, adjusting cultural norms, and investing in the structures that connect operational knowledge to senior decision-making authority. The cost of not addressing it is documented, in case after case, in the disaster literature. The comparison does not require a sophisticated analysis to resolve (Figure 4.1).

Chapter 5

Normalised Tolerance

When Dangerous Becomes Normal

How repeated exposure to unaddressed risk produces a cultural revision of what is acceptable: Space Shuttle Challenger and Space Shuttle Columbia.

The Gradual Revision of the Acceptable

One of the most consistent features of the accounts given by people who were present in organisations before major disasters is a particular kind of retrospective dissonance. When they are asked, after the event, whether they were aware that something was wrong, the answer is almost never a clean no. What they describe, more typically, is something harder to categorise: an awareness that was present but had lost its urgency, a concern that had been felt but had gradually been accommodated into the background of normal professional life, a sense that conditions were not quite right that had been, over time, absorbed into the revised understanding of how the organisation operated.

This is Normalised Tolerance at the level of individual experience. At the organisational level, it is something more structural and more dangerous: the cultural embedding of progressively more permissive understandings of acceptable operational practice, produced by repeated exposure to a known risk that does not, on repeated occasions, produce immediate catastrophic consequences.

Vaughan (1996) called this normalisation of deviance, the process by which a deviation from established safe practice, through its repeated occurrence without catastrophic outcome, is gradually reclassified within the organisational culture as an acceptable characteristic of normal operations rather than a hazard requiring urgent attention. Her account of how this process operated within NASA in the years between the first shuttle flights and the Challenger disaster is the most forensically detailed examination of the phenomenon in the disaster literature, and it remains, nearly three decades

DOI: 10.4324/9781042017539-8

after its publication, the definitive analysis of how organisations learn to tolerate the conditions that eventually destroy them.

What makes the NASA story particularly instructive, and particularly disturbing, is not simply that the normalisation of deviance occurred. It is that it occurred in an organisation composed of highly intelligent, technically sophisticated, genuinely safety-committed professionals. People who were capable of extraordinary analytical rigour in every domain of their work. People who cared deeply about the safety of the astronauts in their care. The normalisation did not happen because these people were negligent or indifferent. It happened because the organisational conditions in which they were working produced, through entirely ordinary mechanisms, a progressive revision of the epistemic standards by which they evaluated risk. By the time of the Challenger launch decision, they were not making a reckless choice. They were making what appeared to them to be a reasonable professional judgement, one shaped by years of accumulated organisational experience that had, without their fully recognising it, revised the boundaries of what counted as acceptable.

This chapter examines that process in depth through two case studies separated by 17 years: Challenger in 1986 and Columbia in 2003. The fact that the same organisation, having experienced the Challenger disaster and conducted a thoroughgoing inquiry into its causes, produced the same pattern of Normalised Tolerance in the years leading to the Columbia disaster is perhaps the most powerful evidence available that normalisation of deviance is not an aberration of particular organisations at particular moments. It is a structural tendency of complex organisations operating under production pressure, and it reasserts itself unless it is actively and continuously interrupted.

Case Study One: Space Shuttle Challenger, 28 January 1986

Case Overview: *On 28 January 1986, the Space Shuttle Challenger broke apart 73 seconds after launch, killing all seven crew members. The immediate cause was the failure of the O-ring seals in the right solid rocket booster field joint under unusually cold launch conditions. The O-ring anomaly had been a known engineering concern since the earliest shuttle flights. Over the preceding nine years, through a process that Vaughan (1996) would document in forensic detail, the anomaly had been transformed, within the organisational culture of NASA and its contractor Morton Thiokol, from a recognised hazard requiring urgent engineering resolution into an accepted operational characteristic of the shuttle programme.*

The Engineering Problem and Its Discovery

The solid rocket boosters used to propel the Space Shuttle from the launch pad were constructed in segments joined by field joints, connections between the cylindrical sections of the booster casing that had to withstand the extreme pressure and temperature of the combustion gases produced during launch. Each joint was sealed by two rubber O-ring seals, designed so that the primary O-ring would provide the main seal against gas escape and the secondary O-ring would provide redundant protection if the primary failed.

From the second shuttle flight in 1981, engineers began observing evidence that the joint design was not performing entirely as intended. The phenomenon they were seeing, which came to be known as blowby, involved the erosion of the primary O-ring seal by hot combustion gases that had breached the putty layer intended to protect it before the primary O-ring seated fully. This was not a trivial finding. The design specification required the primary O-ring to seal the joint. Erosion of the primary seal by hot gases was evidence that the joint was, at least transiently, unsealed and that the secondary O-ring was providing the protection that the design had not intended it to need for routine operations.

The anomaly was observed, documented, and reported. Engineers at both NASA and Morton Thiokol, the contractor responsible for the solid rocket boosters, recognised its significance and raised concerns about what it meant for the safety of continued flight. Analyses were conducted. Recommendations were made. And then, across the following years, something happened that Vaughan's analysis traces with extraordinary precision: the anomaly did not go away, but the organisational response to it changed.

The Mechanics of Normalisation at NASA

Vaughan's central analytical contribution is her account of the specific organisational mechanisms through which the normalisation of the O-ring anomaly occurred. She is at pains to distinguish this process from simple negligence or deliberate risk-taking, and this distinction is what makes her analysis so important for understanding how Normalised Tolerance operates in organisations more generally.

The normalisation did not occur through any single decision. It occurred through an accumulation of decisions, each of which was, in the context of the information and cultural norms available to the people making it, defensible as an exercise of technical judgement. The key mechanism was what Vaughan calls the decision to proceed under uncertainty: the repeated determination, following analysis of the O-ring anomaly data, that the evidence available was sufficient to support a conclusion that the risk was acceptable for the next flight.

This determination was made through NASA's formal Acceptable Risk Process, a structured procedure through which engineering anomalies were assessed and either resolved or accepted as risks within the programme's operating parameters. The process required engineers to demonstrate, through quantitative analysis, that an anomaly did not present an unacceptable threat to mission success. For the O-ring anomaly, this analysis was conducted repeatedly across the years of the programme, each time producing a conclusion that, while the anomaly was present and its cause was not fully understood, the evidence of previous flights supported the judgement that it could be managed within acceptable risk parameters.

Vaughan (1996) traces the epistemic transformation that occurred through this repeated process. In the earliest flights, the absence of a catastrophic outcome following O-ring erosion was treated as ambiguous; it might mean the anomaly was within acceptable limits, or it might simply mean that fortune had favoured the mission. Over subsequent flights, as the anomaly recurred without catastrophic consequence, the accumulated flight data became evidence, evidence that the anomaly was, as operational experience had demonstrated, within the range of what the system could tolerate. The very recurrence that should have intensified concern was reinterpreted, through the logic of the Acceptable Risk Process, as reassurance. The deviance had been normalised not through inattention but through analysis, through the rigorous application of technical expertise within a framework that had itself become structurally biased toward the conclusion that flight was acceptable.

The Temperature Question and Its Suppression

The night before the Challenger launch, engineers at Morton Thiokol raised an explicit objection to the launch on the grounds of temperature.

The forecast overnight temperature at the Kennedy Space Center was significantly below the lowest temperature at which the shuttle had previously launched, and the Thiokol engineers had data suggesting that O-ring resilience decreased with temperature, that cold O-rings were less able to seat quickly against the joint gap under launch conditions and, therefore, more vulnerable to the kind of blowby that had been observed on previous flights.

The teleconference that followed, in which Thiokol engineers presented their data and their recommendation to delay the launch, is one of the most documented and analysed decision-making episodes in the history of safety science. The engineers' data was, in retrospect, clearly pointing toward a dangerous launch condition. The O-ring performance at cold temperatures had been compromised on every previous cold-weather launch. The recommendation not to launch below a temperature significantly higher than the forecast overnight low was grounded in genuine engineering concern.

The recommendation was not accepted. The Thiokol management team, under pressure from NASA managers who challenged the data and the recommendation, reversed the engineers' position and approved the launch. The specific mechanism of that reversal, which involved a request from a senior NASA manager for Thiokol to take off their engineering hats and put on their management hats, has become one of the most cited examples in the organisational safety literature of how production pressure and managerial authority can override technical expertise in the final moments before a catastrophic decision (Vaughan, 1996; Presidential Commission on the Space Shuttle Challenger Accident, 1986).

But Vaughan's deeper point is not about the final teleconference. Her point is that the final teleconference would not have unfolded the way it did if the preceding years had not produced the conditions of Normalised Tolerance that shaped the epistemic context in which it occurred. The Thiokol engineers were arguing against a background of flight history in which the O-ring anomaly had been formally assessed and formally accepted as an operational risk. The NASA managers who challenged their data were not simply being reckless. They were applying the standards of evidence that the organisation's prior analytical processes had established as sufficient. The normalisation that had occurred over years of flight operations shaped the final decision in ways that were invisible to the people making it, because the normalisation was not a departure from the organisation's analytical culture. It was the product of it.

What the Presidential Commission Found

The Presidential Commission on the Space Shuttle Challenger Accident, chaired by former Secretary of State William Rogers, documented the technical causes of the disaster and the decision-making failures of the night before the launch with considerable thoroughness. Its report is a valuable primary source for understanding both the immediate causation of the disaster and the broader organisational context in which it occurred.

The Commission found that the decision to launch Challenger was flawed. It documented the temperature concerns raised by the Thiokol engineers, the pressure applied by NASA managers, and the reversal of the initial no-launch recommendation. It recommended significant changes to NASA's decision-making processes and its safety management systems (Presidential Commission on the Space Shuttle Challenger Accident, 1986).

What the Commission did not fully capture, and what Vaughan's subsequent academic analysis provided, was the deeper organisational history through which the conditions for the launch decision had been produced. The Commission's framework was, in important respects, still shaped by the logic of the blame attribution model: it identified failures of communication, management pressure, and procedural adherence, and it made recommendations addressed to those specific failures. Vaughan's contribution was to show that those failures were themselves the products of the normalisation process, that the communication failures and management pressures of 27 January 1986 could only be understood in the context of the nine years of organisational history that had produced the epistemic and cultural conditions in which they occurred.

The Challenger disaster was not caused by a decision to take an unacceptable risk. It was caused by an organisational process that had, over years of normal operation, redefined the boundary of the acceptable until a risk that should have been clearly unacceptable appeared, to the people responsible for the launch decision, to fall within the bounds of what the organisation had learnt to tolerate.

Case Study Two: Space Shuttle Columbia, 1 February 2003

Case Overview: *On 1 February 2003, the Space Shuttle Columbia disintegrated during re-entry into Earth's atmosphere, killing all seven crew members. The immediate cause was damage to the thermal protection system on the leading edge of the left wing, sustained when a piece of foam insulation from the external tank broke off during launch and struck the wing at high velocity. Foam shedding during launch had been observed on multiple previous flights. By 2003, it had been normalised and accepted within NASA's operational culture as an understood and manageable characteristic of shuttle launches, in a process that the Columbia Accident Investigation Board found to be almost precisely analogous to the normalisation of the O-ring anomaly that had preceded Challenger 17 years earlier.*

Foam Shedding and Its History

External tank foam shedding, the loss of pieces of the foam insulation applied to the shuttle's external fuel tank during the aerodynamic stresses of launch, had been observed on shuttle flights from the earliest years of the programme. Foam loss was, in engineering terms, an understood phenomenon: the foam was a complex, hand-applied thermal insulation system that was vulnerable to manufacturing defects and environmental conditions, and the stresses of launch were sufficient to detach pieces of varying size from the tank surface.

The potential for foam debris to damage the shuttle's thermal protection system, the tiles and reinforced carbon-carbon panels that protected the orbiter from the extreme heat of re-entry, had been recognised as a theoretical concern from the programme's early years. The thermal protection system was, by design, not intended to withstand significant impact during launch. If a piece of foam large enough to cause meaningful damage struck the orbiter at a vulnerable angle and velocity, the consequences during re-entry could be catastrophic.

This was known. It was documented in the engineering literature of the programme. And over the course of the shuttle's operational history, it was progressively normalised, accepted as a feature of shuttle operations that had been observed repeatedly without producing a catastrophic outcome, and therefore treated as within the range of manageable operational risk.

The Columbia Accident Investigation Board and the Mirror Image

The Columbia Accident Investigation Board, which conducted the most thorough and analytically sophisticated inquiry into either shuttle disaster, found that the organisational conditions that produced the Columbia disaster were, in their essential structure, the same conditions that had produced the Challenger disaster 17 years earlier. This finding, that NASA had not, in the intervening years, fundamentally altered the organisational dynamics that the Challenger inquiry had identified, is one of the most significant and most sobering findings in the disaster literature.

The Board found that NASA's organisational culture had, in the years following Challenger, undergone significant formal change, restructured safety programmes, revised management procedures, and adopted new communication protocols without undergoing the deeper cultural change that those formal reforms were intended to produce. The production pressure that had shaped the Challenger launch decision had reasserted itself in the context of the International Space Station programme, which created intense schedule pressure on the shuttle operations that supported it. The normalisation of deviance that had operated on the O-ring anomaly had reasserted itself on the foam-shedding anomaly through essentially the same mechanism: repeated occurrence without catastrophic outcome generating a progressively more permissive assessment of the risk (Columbia Accident Investigation Board, 2003).

The Board documented in detail the specific decisions through which the foam-shedding anomaly had been normalised in the years before Columbia. Foam loss had been observed and recorded across multiple flights. Each instance had been reviewed. Each review had produced a determination that the observed foam loss was within acceptable parameters and that the available evidence supported the conclusion that it did not represent an unacceptable risk to the orbiter. The cumulative history of flights on which foam loss had occurred without catastrophic consequence had become, as it had with the O-ring anomaly, evidence of the anomaly's acceptability rather than evidence of accumulating danger.

The Columbia Mission and the Foam Strike

On 16 January 2003, during the launch of STS-107, a piece of foam approximately the size of a briefcase detached from the bipod ramp area of the external tank 81.7 seconds after launch and struck the leading edge of Columbia's left wing. The impact was captured on launch camera footage and was visible to engineers reviewing the imagery in the days following launch. The foam strike was, by the standards of previous shuttle flights, unusually large. The bipod ramp area was recognised as a historically significant source of foam shedding. The potential for tile damage was real and was raised by engineers who reviewed the footage.

The response within NASA to the concerns raised about the foam strike across the 16 days of the Columbia mission, during which the crew was in orbit and while the wing damage was present but unknown, is one of the most thoroughly documented examples of Normalised Tolerance operating in real time. Engineers who were concerned about the potential damage raised the issue through the organisation's safety and engineering channels. Requests were made for satellite imagery of the wing that might have revealed the extent of the damage. Those requests were ultimately not acted upon through official channels, in part because the management culture that governed the mission had determined, consistent with the accumulated history of foam-shedding events, that the strike was within the range of previously accepted anomalies and did not represent a safety of flight issue (Columbia Accident Investigation Board, 2003).

The Board's analysis of these decisions is damning in its precision. It found that the determination that the foam strike was not a safety of flight issue was not grounded in engineering analysis; there was no engineering analysis available that could have demonstrated this, because no one had modelled the consequences of a bipod ramp foam strike of this size on the thermal protection system. The determination was grounded in a cultural assumption: the assumption, normalised across years of flight operations, that foam shedding was an acceptable operational characteristic of the shuttle. The assumption was wrong. And the cultural conditions that had produced and sustained it, conditions that were, as the Board found, structurally identical to those that had produced the Challenger disaster, were the real cause of Columbia.

The Failure to Learn

The Columbia Accident Investigation Board's finding that NASA had failed to learn the organisational lessons of Challenger is worth dwelling on, because it goes directly to one of the central questions that this book is written to address: why, given the evidence that organisational conditions produce disasters, do those conditions persist and reassert themselves even in organisations that have experienced disaster and conducted serious inquiries into its causes?

The Board offered a specific answer to this question in the context of NASA, and that answer has implications that extend well beyond the space programme. It found that the reforms implemented following Challenger had addressed the visible, formal symptoms of the organisational failure, the communication breakdowns, the management pressure on technical judgement, the inadequacies of specific decision-making processes without addressing the underlying cultural and structural conditions that had produced those symptoms. The safety programme had been restructured. The management hierarchy had been reformed. The communication protocols had been revised. But the production pressure had not been reduced. The epistemic culture, the implicit standards of evidence and the informal norms about whose concerns counted and how uncertainty should be resolved had not been fundamentally changed. And the tendency toward normalisation of deviance had not been interrupted at the level at which it actually operated: the accumulation of operational experience that generated the revised, more permissive understanding of what was acceptable.

Reason (1997) has described this as the problem of safety culture maintenance, the difficulty of sustaining the cultural conditions that support effective safety management in organisations that are also under persistent production pressure. Safety culture, in Reason's account, is not a stable state. It is a dynamic condition that must be actively maintained against the gravitational pull of normalisation. Organisations that reform their safety systems in response to a disaster but do not invest in the continuous, active maintenance of the cultural conditions that those systems require will find, as NASA found, that the conditions for the next disaster reassert themselves through the ordinary operation of the reformed system.

The Normalisation Cycle: How Tolerance Becomes Culture

Drawing on the Challenger and Columbia cases, and on the broader literature on normalisation of deviance, it is possible to describe the characteristic cycle through which Normalised Tolerance develops in organisations. This cycle is not unique to NASA. It operates wherever repeated exposure to a risk that does not immediately produce catastrophic harm generates a progressive revision of the organisational understanding of what constitutes acceptable practice.

Stage One: Initial Recognition

A deviation from established safe practice is observed and documented. It is initially treated as significant, as a hazard requiring investigation, analysis, and resolution. This is the stage at which the anomaly is clearly visible as an anomaly: it is outside the expected parameters of safe operation, it has been identified as such, and it is being treated with the seriousness that its deviation from safe practice warrants.

This stage is important not because the normalisation process is occurring here – it is not yet – but because it establishes the baseline against which subsequent revisions of the acceptable can be measured. The initial recognition of the anomaly as a hazard is the point at which the organisation's understanding of acceptable practice is most clearly aligned with its actual risk state. Everything that follows moves that alignment progressively further apart.

Stage Two: Analysis under Production Pressure

The initial hazard recognition triggers an analytical process: an investigation of the anomaly, an assessment of its significance, and a determination of whether it is safe to continue operations while it persists. This analysis occurs within an organisational context in which production pressure is real and significant. The costs of halting operations, financial, reputational, and competitive, are immediate and tangible. The benefits of the caution that halting would represent are abstract and uncertain.

In this context, the analytical process is not neutral. It occurs within an epistemic culture, a set of shared assumptions about what counts as sufficient evidence, how uncertainty should be resolved, and whose concerns carry analytical weight, that has been shaped by the organisation's operational history and its production pressures. Where that culture is sound, the analysis will produce conservative conclusions: when in doubt, do not proceed. Where the culture has been shaped by production pressure, the analysis will tend toward conclusions that allow operations to continue. The anomaly will be assessed against a standard of evidence that is, in effect, biased toward the conclusion that the risk is acceptable.

Stage Three: Repeated Exposure without Catastrophic Outcome

The determination that the anomaly is within acceptable limits allows operations to continue. The anomaly recurs on subsequent operational cycles.

Each recurrence without a catastrophic outcome generates a new data point that is added to the analysis of the anomaly's acceptability. Over time, the accumulated history of operational cycles in which the anomaly was present and catastrophe did not occur becomes the primary evidence base for the ongoing assessment of the anomaly's risk.

This is the core of the normalisation mechanism. The very fact that the anomaly has not produced catastrophic harm in the past is used as evidence that it is unlikely to produce catastrophic harm in the future. This reasoning, which Vaughan (1996) identifies as a form of what statisticians call the base rate fallacy, ignores the possibility that previous cycles may have been survivable by fortune rather than by the adequacy of the safety margins and that the absence of catastrophic outcomes in the past provides no guarantee of safety in future conditions that may differ in critical ways.

Stage Four: Cultural Embedding

Through repeated cycles of analysis that produce acceptable conclusions, the anomaly becomes culturally embedded as a known and manageable feature of operations. It is no longer, in the shared understanding of the people who work with it, an anomaly at all. It is simply a characteristic of how the system operates, one of many features of operational life that experienced professionals know about, manage around, and have come to regard as within the range of what they can handle.

At this stage, the normalisation is complete in its most important sense. The organisation's culture, its shared assumptions, professional norms, and informal standards of acceptable practice have been revised to incorporate the previously deviant condition as normal. A new member of the organisation, learning from their colleagues what it means to operate professionally within this context, will absorb the normalised understanding as part of their professional socialisation. The historical baseline, the original understanding of the anomaly as a hazard requiring resolution, is no longer the operative standard. It has been replaced by the culturally embedded standard that years of operational experience and organisational analysis have produced.

Stage Five: Resistance to Correction

Once normalisation has reached this stage, once the deviant condition has been culturally embedded as normal, it becomes resistant to correction in ways that are important to understand. The individuals who raise concerns about the condition are no longer challenging a recognised hazard. They are challenging an accepted understanding of how the system works. Their concerns are assessed not against the original safe operating parameters but against the revised cultural standard. They may be dismissed as overcautious, as insufficiently experienced, or as insufficiently understanding of the organisation's operational history. The engineering data they present is evaluated within an interpretive framework that has been shaped by years of normalisation, a framework that is, in effect, constructed to produce reassuring conclusions.

This dynamic was visible in both the Challenger and Columbia cases. In both instances, engineers who raised concerns about the anomaly in question faced exactly this form of resistance. Their data was challenged. Their methodology was questioned. Their conclusions were treated as insufficiently supported by operational experience. They were not dismissed as incompetent; they were treated as failing to adequately account for what the organisation's accumulated experience had established as the appropriate baseline for risk assessment. The normalisation had, at this stage, become self-protecting.

Identifying Normalised Tolerance in Your Organisation

The practical challenge that Normalised Tolerance presents for leaders and safety professionals is its invisibility from inside the organisation. Once normalisation has occurred, the deviant condition is no longer visible as deviant to the people who work within it. Identifying it requires a specific kind of analytical discipline, one that is capable of examining the organisation's current operational practices against an external baseline rather than against the internally generated standard that the normalisation process has produced.

The first indicator of Normalised Tolerance is the prevalence of language that describes known hazards as managed risks. When experienced professionals in an organisation describe a known hazard with the relaxed familiarity of something that has been observed many times without serious consequence, when the language used to discuss it reflects accommodation rather than concern, this is a cultural signal that normalisation may have occurred. The language of normalisation sounds like competence. It sounds like experience. It sounds like the professional confidence of people who know their system and understand its characteristics. What it may actually represent is the cultural embedding of a revised and more permissive understanding of acceptable risk.

The second indicator is the pattern of response to new concerns raised about known anomalies. In organisations where Normalised Tolerance is operating, concerns about conditions that have been previously analysed and accepted will tend to be met with reference to the prior analytical history rather than with fresh analysis. The response to a new concern will be: we know about this; we have assessed it; it is within acceptable parameters. The question that should be asked, has anything changed that might affect whether our prior assessment remains valid? is typically not asked because the cultural embedding of the normalised condition makes its acceptability appear self-evident.

The third indicator is the gap between the organisation's formal safety standards and its actual operational practices. Every organisation that has experienced Normalised Tolerance will show, on examination, a gap between the conditions that its formal safety systems specify as acceptable and the conditions that are actually being tolerated in operational practice. Identifying this gap requires direct examination of operational practice, observation of what is actually happening in the field, rather than a review of formal documentation. The formal documentation will, in most cases, reflect the originally specified standards. The operational practice will reflect the normalised ones.

The fourth indicator, and the one that requires the most leadership courage to act on, is the historical record of how the organisation has responded to anomalies over time. An organisation that can demonstrate a consistent history of treating recurring anomalies as progressively more acceptable, in which the records show a series of analyses, each of which concluded that the same anomaly was within acceptable parameters, each building on the conclusions of the previous one, is demonstrating the signature of Normalised Tolerance in its own documentation. The analytical chain, traced back to its origins, will reveal the baseline from which the organisation has progressively departed. That baseline is the standard against which the current normalised condition should be assessed.

> *Normalised tolerance does not announce itself. It develops through processes that look, from inside the organisation, like rigorous analysis, accumulated expertise, and sound professional judgement. Identifying it requires the willingness to ask a question that organisational cultures shaped by normalisation find profoundly uncomfortable: what if the experience that tells us this is acceptable is the very thing that has made us unable to see that it is not?*

The Responsibility of the Organisation

The analysis of Normalised Tolerance in the NASA cases, and in the disaster literature more broadly, raises a question of organisational responsibility that this book addresses in depth in Chapter 10. But it is worth stating here, because it is often obscured by the focus on individual decision-making that characterises most disaster inquiry processes.

The normalisation of deviance at NASA was not the product of individual incompetence or individual moral failure. It was the product of organisational conditions, production pressure, epistemic culture, and analytical frameworks biased toward acceptable conclusions that were themselves the product of choices made at the organisational and institutional level. The organisation had chosen, repeatedly and structurally, to prioritise schedule and mission success in ways that shaped the conditions in which the normalisation occurred. The individuals who made the final decisions, the managers who approved the Challenger launch, and the managers who determined that the Columbia foam strike was not a safety of flight issue were operating within a context that those organisational choices had produced. Their individual accountability is real. It does not exhaust the accountability of the organisation.

For leaders who want to interrupt the normalisation process rather than simply assign accountability for its consequences, the implications are clear. The conditions that produce Normalised Tolerance, production pressure, epistemic culture biased toward reassuring conclusions, and the absence of mechanisms for fresh analysis of longstanding anomalies are organisational conditions. Addressing them requires organisational decisions: about resource allocation, about the structural independence of safety functions, about the cultural standards through which professional expertise is evaluated, and about the epistemic norms that determine how uncertainty is resolved when operational decisions are made.

These are not easy decisions to make. They require leaders to resist the gravitational pull of production pressure and to invest in safety systems that do not produce immediate, visible returns. They require organisations to create the cultural conditions in which raising concerns about normalised conditions is professionally supported rather than professionally penalised. They require the institutional commitment to treat the absence of a catastrophic outcome as insufficient evidence of safety and to maintain the analytical rigour and the cultural openness that the management of major hazard risk genuinely demands.

The evidence from NASA suggests that these requirements are not beyond the capacity of even large, complex, production-pressured organisations to meet. What the evidence also suggests, with uncomfortable force, is that meeting them requires sustained, active, deliberate effort and that the moment an organisation stops actively maintaining the conditions that support effective safety culture, the normalisation process begins its quiet, invisible, self-reinforcing work.

Chapter 6

Structural Secrecy

When No One Has the Full Picture

How the fragmentation of safety-critical information across organisational boundaries prevents the complete picture of accumulating risk from ever becoming visible, Pike River and Esso Longford.

The Fragmentation Problem

There is a particular quality to the evidence produced by disaster inquiries that deserves more attention than it typically receives. When the inquiry teams reconstruct the pre-disaster history of a major organisational failure, assembling the incident reports, the engineering analyses, the management decisions, the communications, the maintenance records, and the safety audits, what they consistently find is that the information required to understand the developing hazard was, in the aggregate, present within the organisation. Not in any single place, not held by any single person, but distributed across the organisation in fragments, each fragment individually insufficient to constitute a clear warning but collectively forming a picture that is, in retrospect, alarming in its completeness.

This is the phenomenon that Vaughan (1996) called Structural Secrecy. The term is deliberately chosen to distinguish this form of organisational blindness from deliberate concealment. The secrecy is structural rather than intentional: it is produced by the ordinary fragmentation of information in complex organisations rather than by any decision to hide safety-relevant knowledge from the people who need it. The information exists within the system. It is held, in pieces, by different teams in different parts of the organisation. What does not exist is the mechanism, the cross-functional communication, the integrated risk picture, or the shared analytical framework, that would allow those pieces to be assembled into the complete picture that would make the developing hazard visible.

Structural Secrecy is the third condition of the Blindspot Model, and it is in some ways the most insidious of the five conditions because it operates silently and without obvious symptoms. An organisation experiencing

DOI: 10.4324/9781042017539-9

Signal Attenuation knows, at least at the operational level, that safety concerns exist. An organisation experiencing Normalised Tolerance has a history of documented anomalies that can, in principle, be traced and re-evaluated. But an organisation experiencing Structural Secrecy may have no individual within it who possesses the complete picture, not because anyone has failed in their individual responsibility, but because the complete picture has never existed in any one place. The ignorance is genuine, and it is genuine at every level of the organisation, including the levels that bear the most significant responsibility for managing the risk.

This chapter examines Structural Secrecy in depth through two cases that demonstrate its operation with particular clarity: the Pike River mine disaster in New Zealand in 2010 and the Esso Longford gas plant explosion in Victoria, Australia, in 1998. Both are cases in which the post-disaster reconstruction of the organisational history revealed a pattern of fragmented safety-critical information that, assembled, would have constituted a clear and serious warning. Both are cases in which the organisational structures that should have enabled that assembly were absent or inadequate. And both are cases with direct relevance to the Australian and Pacific safety regulatory environment within which this book's primary readership operates.

Case Study One: Pike River Mine Disaster, New Zealand, 19 November 2010

Case Overview: *On 19 November 2010, an explosion in the Pike River Coal mine on the West Coast of New Zealand's South Island killed all 29 men working underground at the time. Three further explosions over the following days made recovery of the bodies impossible. The Royal Commission that investigated the disaster found that Pike River Coal had been operating a mine with serious and escalating safety deficiencies for an extended period before the explosion and that the information about those deficiencies, held in fragments across the organisation, its contractors, and its regulators, had never been assembled into the complete picture that would have made the mine's actual risk state visible to those responsible for managing it.*

The Company and Its Mine

Pike River Coal was a New Zealand listed company that had spent more than a decade developing the Pike River mine, a technically challenging underground coal mine in a geologically complex and methane-rich seam beneath the Paparoa Range on the West Coast of the South Island. The mine had been years in development, had consumed substantial capital, and was, by the time of the disaster, at last beginning to produce coal at commercially significant rates. The pressure to produce, financial, reputational, and institutional, was intense. The company was in a fragile financial position. Its shareholders, its lenders, and the communities that had invested hope and local employment in its success were waiting for it to deliver on the promise of its development phase.

The Royal Commission found that this production pressure had shaped, in pervasive and fundamental ways, the safety culture and safety management systems of Pike River Coal. It found that the company had not, at the time of the disaster, implemented an adequate health and safety management system, that the systems it had in place were inadequate to identify and manage the major hazard risks of underground coal mining in a methane-rich environment, and that this inadequacy had developed over an extended period during which the warning signs of an unsafe operation had been present in multiple parts of the organisation and its regulatory environment without ever being assembled into a coherent risk picture (Royal Commission on the Pike River Coal Mine Tragedy, 2012).

The Fragmentation of Safety Knowledge

What the Royal Commission documented, across thousands of pages of evidence and analysis, was a pattern of fragmented safety knowledge that is among the clearest examples of Structural Secrecy in the Australasian disaster literature. The information about the mine's safety deficiencies was not absent. It was present, held in different forms by different parties across the organisation and its external environment.

The mine's workers knew. The men who worked underground at Pike River knew, from their daily operational experience, that the mine had serious problems. They knew about the methane levels that were frequently elevated beyond safe operating limits. They knew about the ventilation system that was inadequate for the size and complexity of the mine's workings. They knew about the electrical equipment that was not appropriately rated for use in a gassy mine. They knew about the production pressure that shaped management decisions about when to halt operations and when to continue. They knew, in the way that experienced underground workers know things, not necessarily with the formal documentation that would allow their knowledge to be assembled into a risk assessment, but with the immediate operational awareness of people living daily with the consequences of the mine's deficiencies (Macfie, 2013).

The mine's management knew parts of the picture. Individual managers had information about specific deficiencies, about maintenance issues, about equipment concerns about the gap between the mine's ventilation design and its operational reality. But this information was not systematically assembled across the management team into a coherent picture of the mine's overall risk state. The management structure was not designed for that kind of cross-functional risk integration. Individual managers managed their own domains. The information they held stayed within those domains unless there was a specific mechanism for sharing it across them, and the evidence before the Commission established that no such mechanism was effectively operating.

The company's own safety and health systems had generated documentation of hazards and near-misses that, taken together, constituted a serious warning about the mine's risk state. Hazard reports had been filed. Near-miss incidents had been recorded. Maintenance deficiencies had been documented. The Commission found that these documents existed within the organisation's systems but had not been subject to the kind of systematic analysis and escalation that would have revealed their cumulative significance. Each document was a fragment. The fragments had not been assembled.

The Regulatory Dimension

The Structural Secrecy at Pike River was not confined to the company's internal organisation. It extended to the regulatory system that was supposed to provide external oversight of the mine's safety performance.

The Department of Labour, which at the time of the disaster was responsible for the regulation of mine safety in New Zealand, had inspected Pike River in the months before the disaster. The Commission found that those inspections had identified concerns about the mine's safety systems and that improvement notices had been issued. But the Commission also found that the regulatory oversight of Pike River had been, in important respects, inadequate, that the inspection programme had not been sufficiently thorough or sufficiently frequent to provide the level of assurance that the mine's hazard profile warranted; and that the information about the mine's safety deficiencies held by the Department had not been effectively integrated with the information held by the company and by the workers to produce the complete risk picture that the situation required (Royal Commission on the Pike River Coal Mine Tragedy, 2012).

This regulatory dimension of Structural Secrecy is important and frequently underappreciated in disaster analysis. The fragmentation of safety-critical information does not stop at the boundary of the organisation. It extends through the regulatory system, through the industry associations, through the insurance and financial institutions that also have access to risk-relevant information about organisations and their operations. A comprehensive account of Structural Secrecy in high-hazard industries must recognise that the complete picture of organisational risk is rarely held entirely within any single institutional boundary and that creating the mechanisms to assemble it requires engagement not just with the organisation's internal information systems but with the full ecosystem of institutions that hold safety-relevant knowledge about its operations.

What the Commission Found

The Royal Commission's report is one of the most thorough and analytically rigorous examinations of Structural Secrecy in the Pacific disaster literature. Its central finding, that Pike River Coal had been allowed to operate an unsafe mine and that the regulatory system had failed to identify and address the mine's safety deficiencies in time to prevent the disaster, rested on a detailed reconstruction of the fragmented information landscape that characterised the organisation and its regulatory environment.

The Commission found that if the information about the mine's safety deficiencies that was held across the organisation, its workforce, and the regulatory system had been assembled and properly analysed, it would have been clear that the mine was not being operated safely and that urgent corrective action was required. The Commission's report is, in this sense, a map of the complete picture that Structural Secrecy had prevented from ever existing in one place while 29 men were alive to benefit from it.

Macfie's (2013) account of the disaster and its aftermath adds an essential human dimension to the Commission's analytical findings.

Her documentation of the workers' knowledge, their awareness of the mine's deficiencies, their concerns about its safety, and their trust that the company and the regulatory system would manage the risks that they could not manage themselves, is a reminder that Structural Secrecy is not an abstract organisational phenomenon. It is a condition with human consequences of the most serious kind. The men who died at Pike River were not ignorant of the mine's problems. They were victims of an organisational and regulatory failure to assemble the knowledge that they, and others, possessed into a picture that would have required someone with the authority to act on it to act.

The workers at Pike River knew the mine was dangerous. The regulators had documented concerns. The management systems had generated hazard reports. The information was there, distributed across the organisation and its environment. What was absent was the mechanism, and perhaps the will, to bring it together into the picture that would have made inaction impossible.

Case Study Two: Esso Longford Gas Plant Explosion, Victoria, Australia, 25 September 1998

Case Overview: *On 25 September 1998, an explosion and fire at the Esso gas processing plant at Longford in Victoria's Gippsland region killed two workers, injured eight others, and cut gas supply to the state of Victoria for two weeks. The Longford Royal Commission found that the accident was the product of a complex set of organisational failures that had developed over an extended period, including the progressive fragmentation of safety-critical engineering knowledge across organisational boundaries following a major corporate restructuring.*

The Plant and Its History

The Longford gas plant had been operating since 1969 as the onshore processing facility for natural gas and condensate produced from the Gippsland Basin offshore fields. By the late 1990s it was processing virtually all of the natural gas supplied to Victoria, a supply dependency that meant any extended interruption to its operations would have immediate and significant consequences for the state's economy and its population.

The plant had been designed and operated by Esso Australia, a subsidiary of ExxonMobil, and had, across its operating history, developed a substantial body of engineering knowledge about the specific hazards of its processes. That knowledge was embedded in the plant's operating procedures, in the expertise of its engineers, and in the institutional memory of the experienced personnel who had worked with its systems for many years. It was, in the language of this book, the organisation's accumulated understanding of its own risk state, the hard-won knowledge of how its systems behaved under normal and abnormal conditions, and what the boundaries of safe operation looked like in practice.

In the years preceding the disaster, a series of corporate restructuring decisions had, with entirely ordinary organisational logic, fragmented that accumulated knowledge in ways that were not immediately visible as dangerous but that had, by 1998, produced a plant in which the critical knowledge required to manage a specific and known hazard was no longer reliably held by the people responsible for operating the systems that hazard affected.

The Restructuring and Its Knowledge Consequences

In the early 1990s, Esso Australia undertook a significant restructuring of its operations that was consistent with the broader corporate trends of the era: cost reduction, organisational delayering, and the transfer of technical functions from specialist engineering roles to generalist operational roles. As part of this restructuring, the resident engineers who had previously worked on-site at the Longford plant, engineers who possessed detailed technical knowledge of the plant's processes and who had provided immediate technical support to operational decisions, were removed from the site and relocated to a central engineering office in Melbourne.

The Longford Royal Commission found that this restructuring had a specific and critical consequence for the management of one of the plant's key hazard scenarios. The engineering knowledge required to understand and manage the behaviour of the plant's heat exchanger equipment under abnormal temperature conditions, specifically, the risk of metal embrittlement and catastrophic failure that could result from the introduction of very cold hydrocarbon liquid into warm equipment, had been held primarily by the resident engineers. When those engineers left the site, the knowledge they possessed did not transfer effectively to the operational personnel who remained. The training systems that should have ensured operational staff understood the hazard were inadequate. The operating procedures that should have encoded the required response to the abnormal condition were incomplete. The engineering expertise that should have been immediately available to support operational decision-making in abnormal situations was now located in Melbourne, not at the plant (WorkSafe Victoria, 1999).

Hopkins (2000) identifies this as a specific mechanism of Structural Secrecy that operates with particular force in organisations that have undergone significant restructuring: the engineering-operations knowledge gap. When technical expertise is separated from operational environments, whether through geographic relocation, organisational delayering, or the substitution of generalist for specialist roles, the knowledge that was previously integrated in the day-to-day management of operational risk becomes fragmented across an organisational boundary. The operators have the operational access. The engineers have the technical knowledge. Neither has both. And the boundary between them, when it is inadequately bridged, becomes the location at which the complete picture of operational risk ceases to exist.

The Day of the Disaster

On the morning of 25 September 1998, operators at the Longford plant were managing an abnormal situation involving one of the plant's heat exchangers, GP 905, that had become extremely cold following an interruption to the flow of warm lean oil that normally kept it at operating temperature. The situation was unusual. The operators managing it were working without the engineering expertise that would have been immediately available before the restructuring and in the absence of operating procedures that adequately described the response required.

The specific hazard, the risk of catastrophic failure when warm hydrocarbon liquid was reintroduced into the cold exchanger, was not unknown to the organisation. It had been identified in engineering analyses that existed within the company's documentation systems. It was knowledge that the resident engineers who had previously worked on-site would have possessed and could have applied to the situation. But the operators managing the situation on 25 September 1998 did not have access to that knowledge. They were managing an abnormal condition in a complex system with inadequate technical support and incomplete procedural guidance, in circumstances where the organisational restructuring had fragmented the knowledge required to recognise the hazard for what it was and respond to it appropriately.

When lean oil was reintroduced into the cold heat exchanger, the metal, embrittled by the extreme cold, failed catastrophically. The resulting explosion killed two men and injured eight others. The gas supply interruption that followed affected every gas-dependent household and business in Victoria for the next two weeks, an impact that the Royal Commission estimated cost the Victorian economy approximately 1.3 billion dollars (WorkSafe Victoria, 1999).

The Commission's Findings on Knowledge Fragmentation

The Longford Royal Commission, conducted by Justice Dawson and Professor Brooks, produced a detailed and analytically sophisticated account of the role that knowledge fragmentation played in producing the conditions for the disaster. Its findings are directly relevant to the concept of Structural Secrecy as developed in this chapter.

The Commission found that the technical knowledge required to manage the hazard that caused the disaster had existed within the organisation. It had been held by the resident engineers who had worked at the plant prior to the restructuring. It was embedded in engineering analyses that were part of the company's documentation. It was known to personnel at ExxonMobil's international operations who had managed similar hazards in other facilities. What had not existed, since the removal of the resident engineers, was the mechanism to make that knowledge available, in real time, to the operators managing the specific abnormal situation that led to the explosion.

The Commission also found that the training and competency systems that should have transferred the critical knowledge from the departing engineers to the operational personnel who remained had not been adequate to the task. This is a specific and important finding: the knowledge gap produced by the restructuring was foreseeable, and the organisation had a responsibility to address it through training and procedure development that it did not discharge. The Structural Secrecy at Longford was not simply the inevitable consequence of organisational change. It was the consequence of organisational change that was not accompanied by the knowledge management measures required to maintain the integration of engineering expertise and operational practice on which safe operation of the plant depended.

Hopkins (2000) draws a specific lesson from the Longford case that is directly relevant to this chapter's argument: that the decision to remove resident engineers from the plant was, in corporate terms, entirely rational; it reduced costs, it aligned with contemporary management thinking about organisational efficiency, and it was consistent with what other companies in the industry were doing at the time. Its safety consequences were not apparent to the people who made it, because those consequences were a function of the knowledge fragmentation it produced, a fragmentation that was structural, gradual, and invisible from within the corporate decision-making framework that generated it. This is Structural Secrecy at the level of corporate governance: the kind of organisational blindness that produces dangerous conditions through entirely reasonable decisions made by people who cannot see, from where they are sitting, the complete picture of what those decisions are doing to the organisation's risk state.

The Architecture of Structural Secrecy

Drawing on the Pike River and Longford cases, and on the broader literature, three primary architectural features of Structural Secrecy can be identified. Each represents a specific way in which the design of organisations, their structures, processes, and boundaries, produces the fragmentation of safety-critical information that prevents the complete risk picture from becoming visible.

Feature One: Functional Siloing

Functional siloing is the separation of safety-critical information across the functional boundaries of organisational departments, disciplines, and specialisations. In most large organisations, different aspects of the organisation's risk state are managed by different functional groups: engineering, operations, maintenance, safety, human resources, and finance, each of which has access to information relevant to the overall risk picture but none of which has access to all of it.

Functional siloing is not inherently pathological. The specialisation of functions in complex organisations is a rational response to the complexity of the information and expertise required to manage them. The problem arises when the boundaries between functions become barriers to the flow of safety-critical information, when the information relevant to a developing hazard is held in multiple silos without any effective mechanism for its assembly and integration. In this condition, each functional group can be operating entirely appropriately within its own domain while the overall risk picture remains invisible because it exists only in the aggregate of those domains, not in any single place within the organisation.

Feature Two: Geographic and Hierarchical Distance

Geographic and hierarchical distance amplifies functional siloing by adding physical and organisational separation to the barriers that prevent the assembly of fragmented information. At Longford, the removal of resident engineers from the plant site created exactly this kind of distance: the engineering knowledge that had previously been integrated with operational practice at the point of need was relocated to a different city, separated from the operational environment by both geography and organisational hierarchy.

Hierarchical distance operates in a related but distinct way. As information moves upward through the layers of an organisational hierarchy, it is aggregated and summarised at each level, and with each aggregation the specific operational detail that might reveal a developing hazard is progressively lost. The senior executive reviewing a monthly safety performance report is seeing a highly aggregated picture of the organisation's risk state, one that may accurately represent the performance of the formal safety management system while being entirely silent about the operational-level realities that the system is not capturing. The hierarchical distance between that executive and the operational environment has produced a form of Structural Secrecy that is, from the executive's perspective, complete: they have a picture of the organisation's risk state, and it is accurate as far as it goes, but it does not go anywhere near far enough.

Feature Three: System and Process Fragmentation

System and process fragmentation is the distribution of safety-critical information across multiple recording, reporting, and management systems that are not integrated with each other. In complex organisations, information about hazards, near-misses, maintenance deferrals, procedural non-compliances, and safety culture concerns may be held in different systems: a hazard register, an incident reporting database, a maintenance management system, a safety audit tracking system, and a contractor management system, none of which communicates with the others and none of which, therefore, can produce the integrated risk picture that the complete information would support.

This form of Structural Secrecy has become, if anything, more prevalent with the growth of digital information management systems in the past two decades. Organisations now generate and store more safety-relevant data than at any previous point in history. The challenge is not the volume of data. The challenge is the integration of that data into a picture of the organisation's actual risk state that is meaningful, current, and actionable. Organisations that have invested heavily in safety management software without investing equally in the analytical processes and human judgement required to make sense of the data those systems generate may have more Structural Secrecy than they had before the investment, more data distributed across more systems and less capacity to assemble it into the complete picture that the data, taken together, would reveal.

Weick and the Problem of Collective Sensemaking

The concept of Structural Secrecy connects directly to one of the most important theoretical frameworks in the organisational behaviour literature for understanding how organisations fail to see what is happening within them: Weick's account of sensemaking and its failure.

Weick (1995) argues that organisations do not simply collect and process information; they actively construct their understanding of their environment through ongoing processes of social sensemaking: the collaborative, culturally embedded practices through which organisational members interpret the ambiguous and fragmented information available to them and produce shared understandings of what is happening and what it means. Sensemaking in organisations is inherently a collective process. It depends on the flow of information between people, on the quality of the communication structures that enable that flow, and on the cultural norms that shape how information is interpreted and what gets made of it.

Structural Secrecy attacks the sensemaking process at its foundations. When safety-critical information is fragmented across organisational boundaries, the collective sensemaking that would assemble it into a coherent risk picture cannot occur, not because the people involved are incapable of it, but because the information required for it does not flow to the places where it could be made sense of. The individual pieces are available. The process of collective interpretation that would reveal their significance is not present.

Weick et al. (1999) developed this insight in their account of high-reliability organisations, arguing that the organisations that consistently manage complex, high-hazard systems with the lowest rates of catastrophic failure are those that have developed specific organisational practices for maintaining collective mindfulness: the shared, active, and ongoing awareness of the organisation's current operational state, including its deviations from intended performance and the significance of those deviations for the overall risk picture. Collective mindfulness is, in effect, the organisational antidote to Structural Secrecy – the capacity to assemble fragmented information into a coherent picture in real time – before that fragmentation allows a developing hazard to reach the point of catastrophic failure.

Identifying and Addressing Structural Secrecy

Structural Secrecy is, by its nature, difficult to identify from inside the organisation, because its defining characteristic is the absence of the complete picture, and the absence of something is inherently harder to notice than the presence of something. The practical challenge for leaders and safety professionals is to develop the organisational habits of inquiry that can detect the signature of fragmented information before the consequences of that fragmentation become catastrophic.

The first diagnostic question is: can anyone in this organisation, at any level, describe accurately and completely the current major hazard risk profile of our operations? If the answer is no, if producing a complete and current picture of the organisation's major hazard risk state would require assembling information from multiple systems, multiple functions, multiple sites, and multiple levels of the hierarchy, then Structural Secrecy is present. The question is not whether the individual fragments of that picture exist within the organisation, but whether they can be assembled into a coherent whole by any individual or team with the authority to act on what the complete picture reveals.

The second diagnostic question is, what happened to the technical and operational knowledge that was held by people who have left the organisation, been transferred to different roles, or been replaced by contractors? Corporate restructurings, senior personnel changes, and the substitution of contractor labour for direct employment are among the most common sources of the engineering-operations knowledge gap that the Longford case illustrates. Organisations that cannot account for where critical operational knowledge has gone following significant structural or personnel changes are organisations in which Structural Secrecy may be developing in ways that are not yet visible.

The third diagnostic question is, what information reaches the senior leadership team's regular review of safety performance, and what does not? A systematic comparison of the information available in the formal safety performance reporting with the information available at the operational level, through direct worker consultation, through safety audits, and through independent technical assessments, will typically reveal the gap that Structural Secrecy has produced. The gap itself is the finding. Its size and character will indicate where the fragmentation is occurring and what organisational mechanisms are needed to address it.

The fourth diagnostic question, and the one that requires the most sustained leadership attention, is, who has the responsibility and the authority to assemble the complete picture? In most organisations experiencing Structural Secrecy, the answer is either nobody or nobody whose authority is sufficient to act on what the complete picture reveals. Addressing Structural Secrecy requires not just better information systems but also the deliberate organisational investment in the roles, the processes, and the cultural norms that make the assembly of fragmented safety information both possible and consequential and that ensure the complete picture, when it is assembled, reaches the people who can and will act on it.

> *Structural Secrecy is not a failure of information collection. Most high-hazard organisations collect more safety-relevant information than they know what to do with. It is a failure of information integration, the failure to create the organisational conditions in which the fragments of the risk picture that exist in different parts of the system can be brought together into the complete picture that would make the developing hazard visible. That failure is organisationally produced. It is organisationally addressable.*

The Relationship Between Structural Secrecy and the Other Conditions

Structural Secrecy does not operate in isolation. Its relationship to the other four conditions of the Blindspot Model is important for understanding both how it develops and how it can be most effectively addressed.

The relationship between Structural Secrecy and Signal Attenuation is particularly close. Both conditions involve the failure of safety-critical information to reach the decision-makers who need it. But they operate through different mechanisms. Signal Attenuation involves the weakening of information that is moving upward through the organisation, information that exists in a form capable of transmission but that is filtered and softened before it arrives. Structural Secrecy involves the failure of information to be assembled in the first place, the fragmentation of the risk picture at the point of origin before any question of upward transmission arises. An organisation can have effective upward communication channels and still be experiencing Structural Secrecy if the information that enters those channels is already incomplete because it has never been assembled from its fragmented sources.

The relationship between Structural Secrecy and Leadership Detachment is equally significant. Leaders who are detached from operational reality are, by definition, working with an incomplete picture of their organisation's risk state. But not all of that incompleteness is the product of attenuation through the reporting hierarchy. Some of it is the product of Structural Secrecy: the information that the complete picture would provide simply does not exist in any assembled form within the organisation. When leaders try to close the gap between their understanding of the organisation and its operational reality, as mindful leaders must, they may find that the complete picture they are seeking cannot be provided not because of cultural resistance or hierarchical filtering but because it has never been assembled. Addressing that problem requires the organisational investments in integration that this chapter has described.

Structural Secrecy also amplifies the effects of Normalised Tolerance. When the complete picture of a developing anomaly's history is fragmented across multiple systems and functions, the analysis of that history that the normalisation process requires, the accumulation of operational experience that generates the revised, more permissive understanding of acceptable risk, tends to be conducted on incomplete data. Each analytical step in the normalisation cycle is taken without the benefit of the full historical record of the anomaly's development and the full range of concerns that have been raised about it. The normalisation occurs not just despite adequate information but in the absence of it, on the basis of a partial picture that has been shaped as much by what the Structural Secrecy has hidden as by what it has revealed.

These interactions underscore the argument of Chapter 3: that the five conditions of the Blindspot Model are not independent risk factors but interlocking components of a system that, when all five are operating, produces a form of organisational blindness that is greater than the sum of its parts. Addressing Structural Secrecy in isolation is valuable. Addressing it as part of a comprehensive assessment of all five conditions is what the evidence demands.

Chapter 7

Production Gravity

When Getting the Job Done Becomes the Only Thing

How the relentless pull of economic and operational pressure erodes safety margins, one reasonable decision at a time, BP Texas City and Wittenoom.

The Constant Pull

Every organisation that produces something operates within a gravitational field. The field exerts a continuous pull towards production, towards the delivery of output, the completion of tasks, the meeting of targets, and the satisfaction of the economic imperatives that justify the organisation's existence. This pull does not require a specific actor or a specific decision to operate. It is the background condition of organisational life in any enterprise that must produce in order to survive, and it shapes, at the margin, every decision made within the organisation about the allocation of resources, the management of time, and the prioritisation of competing demands.

Safety management exists within this gravitational field. Safety systems, safety investments, and safety decisions are made by the same people, in the same organisations, under the same pressures as every other operational decision. When the gravitational pull of production is weak, when organisations are not under significant financial or operational pressure, the balance between production and safety can be maintained relatively easily. When the gravitational pull intensifies, when competitive pressure is high, when budgets are under sustained constraint, when operational targets are ambitious and the consequences of missing them are severe, the balance shifts. Not through a single dramatic decision, but through the accumulation of individually reasonable adjustments: the maintenance item is deferred because the equipment is still running, the procedure shortcut is accepted because the schedule does not allow time for the full process, and the safety investment is postponed because this quarter's results require the capital elsewhere.

Dekker (2011) describes this as drift: the gradual, incremental migration of the organisation's operating point toward the boundary of safe performance,

DOI: 10.4324/9781042017539-10

driven not by dysfunction or indifference but by the ordinary optimisation of a system operating under production pressure. Each step in the drift is locally rational; it makes sense in the immediate context of the decision. The cumulative effect of many such steps is an organisation operating dangerously close to its limits, without any single person having made a decision that, in isolation, appeared unreasonable.

This chapter calls that cumulative effect Production Gravity, and it examines its operation through two case studies that are, in important respects, opposites, not in the pattern they exhibit but in the scale, duration, and industry context in which that pattern operated. BP Texas City is a corporate industrial disaster of the early twenty-first century, produced by a decade of cost-reduction culture in one of the world's largest oil companies. Wittenoom is an industrial public health catastrophe of the mid-twentieth century, produced by decades of commercial prioritisation in an asbestos mining operation in Western Australia's remote Pilbara region. The timeframes are different; the industries are different; and the regulatory contexts are different. The gravitational mechanism is the same.

Case Study One: BP Texas City Refinery, Texas, USA, 23 March 2005

Case Overview: *On 23 March 2005, an explosion and fire at BP's Texas City refinery killed 15 workers and injured 180 others in what remains the deadliest industrial accident in the United States in a generation. The immediate cause was the overfilling of a raffinate splitter tower during a restart operation, leading to a release of flammable hydrocarbons that ignited. The underlying causes, documented in exhaustive detail by Hopkins (2008) and the Baker Panel (2007), were a decade-long programme of cost reduction that had systematically degraded the site's process safety management systems, its equipment integrity, and its safety culture.*

BP's Cost Reduction Programme and Its Safety Consequences

BP Texas City was not a poorly managed refinery by the conventional measures of the early 2000s. Its personal safety record, the metric that most organisations and most regulators used to assess safety performance in that era, was, by industry standards, respectable. Its rates of lost-time injuries and recordable incidents had been declining.

Visitors to the site would have seen a safety culture that, in its visible manifestations – safety signage, personal protective equipment compliance, toolbox talks, and safety inductions – appeared to meet contemporary expectations.

What was not visible in those conventional metrics was the state of the site's process safety management: the systems and practices responsible for managing the risk of the major hazard events, fires, explosions, and toxic releases that are the primary life-safety concerns of refinery operations. Process safety performance is not well captured by personal injury rates. A site can have excellent personal safety performance, few slips, trips, and minor injuries, while its major hazard management systems are in a state of serious deterioration. Texas City was exactly such a site. Hopkins (2008) documents this with considerable precision, drawing on the findings of the Chemical Safety Board investigation, the Baker Panel review, and other post-disaster analyses.

The deterioration of Texas City's process safety management had its origins in a sustained programme of cost reduction that BP had implemented across its US refinery operations in the late 1990s and early 2000s. The programme was, in corporate terms, entirely rational: refinery margins were under pressure, the competitive environment was demanding, and reducing the cost base of the refinery operations was a legitimate and important business objective. The decisions that implemented it were made by competent professionals acting within the ordinary constraints of their corporate roles.

The specific consequences of those decisions for process safety management at Texas City were not invisible. Hopkins (2008) documents a series of internal audits, risk assessments, and safety reviews conducted in the years before the disaster that identified serious deficiencies in the site's process safety management systems, deficiencies in equipment integrity, in operating procedures, in training and competency management, and in the management of safet-critical elements. These findings were documented. They were reported through the organisation's formal systems. They were not adequately addressed because the resources required to address them had been substantially reduced by the cost-reduction programme and because the management culture that the cost-reduction programme had produced did not generate the priority for process safety investment that the documented deficiencies required.

The Isomerisation Unit and Its History

The explosion on 23 March 2005 occurred at the raffinate splitter tower in the refinery's isomerisation unit, one of several process units on the site that had been identified in internal safety reviews as having significant process safety deficiencies. The specific deficiencies in the isomerisation unit that contributed to the disaster included: atmospheric blowdown drums and stacks that had been flagged as non-compliant with contemporary process safety standards for many years; trailers housing contract workers located in close proximity to the blowdown stacks in apparent contravention of the site's own siting guidelines; and an instrument and control system that provided inadequate visibility of the liquid level in the raffinate splitter tower during start-up operations.

None of these deficiencies was unknown to the organisation. The atmospheric blowdown systems had been identified as a safety concern in multiple internal and external reviews. The trailer placement had been the subject of internal discussion. The instrumentation limitations were known to the operators who worked with the system. What had not occurred, across the years during which these deficiencies had been documented, was the corrective action that would have been required to address them. The cost of that corrective action, in capital expenditure, in operational disruption, and in the management time and organisational priority required to execute it, had consistently been assessed, in the context of the site's cost-reduction culture, as a lower priority than the production demands of the operation (Baker, 2007).

On the morning of 23 March 2005, operators were restarting the raffinate splitter tower following a maintenance outage. The restart procedure required the tower to be filled with raffinate to a specified level before heating commenced. Instrumentation failures and procedural inadequacies meant that the tower was filled significantly beyond its intended level without the operators being aware of how high the liquid had risen. When the tower overflowed, flammable raffinate vapour was released through the atmospheric blowdown stack. The vapour cloud ignited. The explosion destroyed the trailers housing the contract workers located nearby. Fifteen people died.

The Baker Panel and Its Findings

In the aftermath of the disaster, BP commissioned an independent review of its safety culture and safety management systems across its

US refinery operations, chaired by former Secretary of State James Baker. The Baker Panel's report, published in January 2007, is one of the most significant corporate safety reviews produced in the aftermath of a major industrial disaster, and its findings are directly relevant to the concept of Production Gravity as developed in this chapter.

The panel found that BP had not provided effective process safety leadership and had not adequately established process safety as a core value across its US refinery operations. It found that BP had inappropriately relied on personal injury metrics as a proxy for process safety performance and that the organisation's safety management systems were orientated towards measuring and improving personal safety performance while providing inadequate visibility of the process safety risks that were, in the major hazard context of refinery operations, the primary source of catastrophic risk.

Most significantly for the Production Gravity argument, the panel found that BP's corporate cost-reduction culture had produced, across its US refinery operations, a systematic tendency to underfund process safety management relative to the requirements of safe operation. The panel documented a pattern in which budget pressures consistently resulted in the deferral or elimination of process safety investments, equipment integrity, safety management systems, training and competency that the sites' own internal assessments had identified as necessary. This pattern was not the product of individual managerial negligence. It was the systemic output of a corporate culture and a budget management process that had, in effect, made Production Gravity the dominant force in every resource allocation decision affecting process safety.

Hopkins (2008) draws a specific and important distinction in his analysis of the Texas City disaster between what he calls individual safety, the management of risks to individual workers from the ordinary hazards of their work, and process safety, the management of the major hazard risks inherent in the processing of flammable, toxic, or explosive materials. He argues that BP's safety culture had been optimised for individual safety performance, producing the personal injury metrics that looked respectable in external assessments, while process safety had been systematically neglected in ways that were invisible to those conventional metrics. The Production Gravity that operated at Texas City was not operating on individual safety; those decisions were visible and measurable and being managed. It was operating on process safety, where the consequences of decisions were deferred, cumulative, and ultimately catastrophic.

The Texas City disaster was not caused by a single bad decision. It was caused by a decade of budget decisions, each individually justifiable, that collectively stripped the site's process safety systems of the resources they needed to function. No one person decided to make Texas City unsafe. The Production Gravity of the cost-reduction culture did that, one capital deferral at a time.

The Metrics Trap

One of the most important lessons of the Texas City disaster, and one that has been widely cited in the subsequent safety literature, is the danger of using personal injury metrics as the primary measure of process safety performance. Muller (2018) has analysed this phenomenon across a wide range of organisational contexts, arguing that the reliance on measurable proxies for the performance of complex systems tends, over time, to optimise the proxy at the expense of the system it is supposed to represent. At Texas City, the proxy, personal injury rates, had been optimised so effectively that it had entirely obscured the state of the system it was supposed to represent, process safety performance.

This is Production Gravity operating through the information system rather than directly through resource allocation. When the metrics an organisation uses to assess its safety performance are measuring the wrong things, the Production Gravity that shapes resource allocation decisions operates without adequate feedback. The organisation's leaders believe, because the metrics tell them so, that safety performance is sound. The Production Gravity that is degrading the unmeasured dimensions of safety performance, process safety management, equipment integrity, and major hazard risk management continues to operate unchallenged because the feedback loop that would reveal its consequences is not functioning. By the time the consequences become visible, they are catastrophic and irreversible.

Case Study Two: Wittenoom, Western Australia, 1943–1966

Case Overview: *The Wittenoom asbestos disaster is Australia's worst industrial public health catastrophe. Between the early 1940s and 1966, Colonial Sugar Refining Company (CSR) and its subsidiary Australian Blue*

Asbestos (ABA) mined and milled crocidolite asbestos at Wittenoom Gorge in Western Australia's Pilbara region under conditions that exposed workers and residents to lethal concentrations of asbestos dust. More than 2,000 people who lived or worked at Wittenoom have since died of asbestos-related diseases, including mesothelioma, asbestosis, and lung cancer. The Production Gravity that shaped the company's management of the asbestos hazard over more than two decades is among the most extensively documented examples of corporate prioritisation of commercial interests over worker safety in Australian industrial history.

The Known Hazard

The health hazards of asbestos exposure were not unknown in the mid-twentieth century. The association between asbestos dust inhalation and serious lung disease, asbestosis, had been established in the medical literature since the 1930s. The specific dangers of crocidolite, the blue asbestos mined at Wittenoom, had been documented. By the time CSR and ABA were operating at peak production in the 1950s, the medical and scientific evidence that exposure to asbestos dust at the concentrations present in the Wittenoom mine and mill posed a serious and potentially fatal health risk to workers was available to the company's management and to the relevant regulatory authorities.

The hazard was known. The question that the Wittenoom story poses, the question that makes it so important for understanding Production Gravity, is not whether the company knew that asbestos dust was dangerous. The question is, what did it do with that knowledge, and why?

The answer, documented in a substantial body of historical scholarship, parliamentary inquiry evidence, and legal proceedings, is that CSR and ABA took the minimum action consistent with the minimum regulatory requirement, deferring and resisting more substantial dust control measures for as long as commercially possible, while continuing to operate a mine and mill in conditions that the company's own dust measurements repeatedly established as far in excess of any credible safe exposure standard.

The Commercial Logic and Its Human Cost

The Wittenoom operation was, by the standards of its time and place, a significant commercial enterprise. Crocidolite asbestos was a valuable industrial material, in high demand for its heat-resistant and insulating properties across a range of manufacturing and construction applications. Australia was one of the world's major sources. The Wittenoom operation was profitable, and its profitability was directly dependent on the cost structure of the mining and milling operation, on the wages paid to workers, the capital invested in infrastructure and equipment, and the operating costs of maintaining the mine.

Effective dust control, the primary technological means of reducing asbestos fibre concentrations in the mine and mill, required capital investment and operational management that would have increased costs and reduced output. Ventilation systems, water suppression, enclosed cabs on mining equipment, respiratory protective equipment, and regular environmental monitoring all cost money. In the commercial logic of the Wittenoom operation, that money was consistently assessed as a cost that could not be justified, or could be deferred, or was not required at the level that an honest engagement with the health evidence would have demanded.

This is Production Gravity in its most nakedly commercial form. The gravitational pull of profitability, of the commercial imperative to minimise cost and maximise output, operated directly on the decisions about dust control. Each decision not to invest in more effective dust control, each decision to maintain production rates in conditions of high dust exposure, and each decision to resist or defer regulatory requirements for improved conditions were decisions shaped by the commercial gravity of an operation whose profitability depended on keeping those costs down. The workers who breathed the dust were not the decision-makers. They were the people on whom the consequences of those decisions landed.

The Regulatory Failure and Its Commercial Facilitation

The Production Gravity at Wittenoom did not operate in a regulatory vacuum. The Western Australian regulatory framework that governed the conditions of the Wittenoom operation was, for the period of the mine's operation, consistently inadequate to the task of compelling the company to manage the asbestos hazard at a level commensurate with what the health evidence required. Dust standards were set too high. Inspections were too infrequent and too deferential. The regulatory response to the company's own dust measurements, which repeatedly and consistently showed concentrations far in excess of even the inadequate standards of the time, was insufficient to produce the corrective action that the measurements demanded.

The regulatory failure at Wittenoom is not unrelated to the Production Gravity that shaped the company's own management decisions. Regulatory systems that are designed and resourced primarily to facilitate production, that treat the economic interests of the regulated industry as a primary consideration in the setting and enforcement of safety standards, are regulatory systems that have themselves been shaped by Production Gravity. The under-regulation of the Wittenoom operation was not accidental. It was the product of a regulatory culture and a political economy that prioritised the economic value of the Wittenoom operation over the health of the people it employed.

This dimension of the Wittenoom story is important for the broader argument of this chapter because it illustrates a point that applies to many of the cases examined in this book: Production Gravity does not operate only within the organisation. It operates across the institutional ecosystem, the regulatory agencies, the industry associations, the political structures, and the financial institutions within which the organisation is embedded. Addressing Production Gravity as an organisational condition without addressing the wider institutional environment that amplifies or constrains it is an incomplete response to the problem the evidence identifies.

The Long Incubation and Its Consequences

What distinguishes the Wittenoom case from most of the industrial disasters examined in this book is its temporal scale. The incubation period at Wittenoom was not years but decades. The Production Gravity that shaped the company's management of the asbestos hazard operated across more than 20 years of mining and milling operations, during which thousands of workers and their families were exposed to lethal concentrations of asbestos fibre in conditions that the company's own measurements consistently documented as dangerous.

The diseases that result from asbestos exposure, mesothelioma, asbestosis, and lung cancer, have latency periods of decades. Many of the men and women who worked at Wittenoom in the 1950s did not develop their asbestos-related diseases until the 1970s, 1980s, or 1990s. The full human cost of the Wittenoom operation, more than 2,000 deaths, with more continuing to occur, was not visible at the time the decisions that produced it were being made. This temporal displacement between the Production Gravity decisions and their human consequences is itself a feature of Production Gravity that the Texas City case, with its immediate catastrophic outcome, does not fully illustrate.

When the consequences of Production Gravity decisions are immediate, when the explosion occurs, when the people die, there is at least the possibility that the causal connection between the decisions and the outcomes will be visible and will produce a corrective response. When those consequences are deferred by decades, the causal chain is obscured. The people who made the decisions that produced the harm are often no longer in the organisation, or no longer alive, by the time the consequences of those decisions become clear. The Production Gravity that shaped the original decisions has no natural feedback mechanism, no explosion, and no immediate catastrophic outcome to interrupt it. It operates unchallenged until the long-deferred consequences eventually make it impossible to ignore.

The Mechanics of Production Gravity

Drawing on the Texas City and Wittenoom cases, and on the broader literature on production pressure and safety management, the specific mechanisms through which Production Gravity operates in organisations can be identified. Understanding these mechanisms is essential for leaders who want to assess whether Production Gravity is operating in their own organisations and what can be done to counteract it.

Mechanism One: Budget-Driven Safety Degradation

The most direct mechanism of Production Gravity is the one documented most thoroughly at Texas City: the systematic under-resourcing of safety management functions through budget processes that prioritise production cost reduction over safety system maintenance. This mechanism operates through the ordinary functioning of corporate financial management, through the annual budget cycle, through capital allocation decisions, through the setting of targets and the management of performance against them.

Budget-driven safety degradation is particularly insidious because each individual budget decision is defensible. The maintenance item deferred this year can be scheduled for next year. The safety investment postponed in this budget cycle can be reprioritised in the next. The training programme reduced to meet a cost target can be supplemented by on-the-job learning. Each of these decisions, in isolation, appears manageable. Their cumulative effect, across multiple budget cycles, across multiple safety management functions, across an extended period of sustained cost pressure, is the systematic degradation of the safety management infrastructure on which safe operation depends.

Mechanism Two: Schedule Pressure and Procedural Erosion

The second mechanism is schedule pressure: the operational urgency that shapes day-to-day decisions about how work is performed and whether safety procedures are followed in full. Schedule pressure does not typically produce decisions to ignore safety procedures entirely. It produces decisions to take shortcuts, to abbreviate procedures, to skip steps that seem unnecessary in the immediate context, and to accept slightly elevated risk in order to meet a deadline.

These shortcuts become, over time, the new operational norm. The abbreviated procedure becomes the way the job is done. The step that was skipped once becomes a step that is routinely skipped. The elevated risk that was accepted as a temporary accommodation to schedule pressure becomes the accepted operational standard. This is the interaction between Production Gravity and Normalised Tolerance: the schedule pressure that produces the initial procedural shortcut creates the exposure without immediate catastrophic outcome that drives the normalisation of the shortcut as acceptable practice. Production Gravity and Normalised Tolerance, in this dynamic, are mutually reinforcing.

Mechanism Three: The Conflation of Personal Safety and Process Safety

The third mechanism, illustrated with particular clarity at Texas City, is the conflation of personal safety performance with overall safety performance in organisations that are under production pressure to demonstrate safety improvement while minimising the costs of process safety management.

Personal safety performance, the management of slips, trips, minor injuries, and other individual harm events, is amenable to relatively low-cost improvement through behaviour-based safety programmes, personal protective equipment compliance, and supervisor engagement with individual safety behaviours. Process safety performance, the management of major hazard risks, requires capital investment in equipment integrity, process control systems, and the major hazard management infrastructure that protects against catastrophic events. Under Production Gravity, the tendency is to invest in the cheaper and more visible personal safety improvements while deferring the more expensive and less visible process safety investments. The resulting safety performance metrics look good. The underlying major hazard risk profile deteriorates.

Muller's (2018) analysis of metric-driven management is directly applicable here. When the measure of safety performance is personal injury rates, organisations will optimise personal injury rates, and the optimisation of the measure will occur at the expense of the underlying reality the measure was intended to represent. The Production Gravity that makes process safety investment expensive also makes personal safety performance improvement attractive. The two tendencies together produce exactly the pattern documented at Texas City: excellent personal safety metrics and catastrophic process safety failure.

Mechanism Four: Commercial Framing of Safety Decisions

The fourth mechanism, most clearly illustrated at Wittenoom, is the explicit commercial framing of safety decisions: the treatment of safety investments and safety standards as cost items to be managed against production economics rather than as non-negotiable requirements of responsible operation.

Commercial framing of safety decisions does not always look like the blatant commercial prioritisation visible in the Wittenoom record. In contemporary organisations, it more typically appears as the requirement to make a business case for safety investments, to demonstrate return on investment for safety programmes, or to assess the cost-benefit ratio of safety measures against their expected risk reduction benefits. These analytical frameworks are not inherently inappropriate; resource allocation decisions in any organisation require some framework for prioritisation. The problem arises when the commercial framework is allowed to operate on safety decisions that are, properly understood, not subject to commercial optimisation: decisions about whether to maintain equipment to the standard required for safe operation, whether to implement dust control measures required to protect workers from known hazards, or whether to invest in the process safety management infrastructure on which safe operation of major hazard facilities depends.

When safety decisions are framed commercially, when the question is not whether to maintain safe operating conditions but whether the cost of maintaining them is justified by the expected risk reduction, Production Gravity has entered the decision-making process through the language and framework of management rather than through the blunt instrument of budget cuts. The gravitational pull is the same. The mechanism through which it operates is more sophisticated and, therefore, more difficult to challenge.

Resisting Production Gravity

The evidence from both the Texas City and Wittenoom cases, and from the disaster literature more broadly, establishes that Production Gravity is a structural feature of organisations operating under production pressure. It cannot be eliminated by good intentions, by safety commitments, or by the individual moral qualities of the leaders responsible for managing it. It can only be managed by deliberate organisational countermeasures: structural, cultural, and systemic features of the organisation that create a genuine counterforce to the gravitational pull of production.

The most important of these countermeasures is the structural independence of safety functions from production management. At Texas City, the process safety management systems that should have provided a check on the cost-reduction culture were located within the same management hierarchy that was driving the cost reduction. They did not have the structural independence, the separate reporting lines, the protected budgets, or the direct access to senior leadership that would have been required to provide a genuine counterweight to Production Gravity. Hopkins (2019) argues that the structural placement of safety functions within organisations is not a matter of administrative convenience. It is a safety-critical decision that determines whether those functions have the authority, the resources, and the independence to perform their essential role as a check on the production pressures that otherwise dominate organisational decision-making.

The second countermeasure is the explicit separation of process safety metrics from personal safety metrics in the organisation's safety performance management systems. The Texas City case demonstrated with fatal clarity what happens when these two dimensions of safety performance are conflated. Organisations that manage major hazard risk must have specific, meaningful leading indicators of process safety performance, indicators that reflect the actual state of the major hazard management infrastructure rather than the personal injury rates that Production Gravity finds it easy and cheap to improve.

The third countermeasure is the explicit acknowledgement, at the senior leadership level, that Production Gravity exists and that resisting it requires active and sustained effort. This acknowledgement is harder than it sounds. Production Gravity is the direction in which the organisation naturally moves. Leaders who acknowledge that they must actively resist it are, in effect, acknowledging that their organisation's natural tendencies, if unchecked, will move it in a dangerous direction. That acknowledgement requires intellectual honesty and institutional courage. It is, however, the necessary foundation of any credible organisational commitment to managing major hazard risk in the presence of significant production pressure.

> *Production Gravity is not a corporate failure. It is a corporate reality. Every organisation that produces under pressure experiences it. The difference between the organisations that manage it and the ones that are eventually destroyed by it is not the presence or absence of the pressure. It is the presence or absence of the deliberate, structural, sustained effort to counteract it before it erodes the margins that separate the organisation from catastrophic harm.*

The Ethical Dimension

Both cases examined in this chapter raise ethical questions that go beyond the analytical framework of the Blindspot Model and that are addressed more directly in Chapter 10. But they deserve brief acknowledgement here, because the ethical dimension of Production Gravity is an important part of what makes it a condition of the Blindspot Model rather than simply a description of normal commercial behaviour.

At Texas City, the Production Gravity that degraded process safety management resulted in the deaths of 15 people who had gone to work that day in the reasonable expectation that the organisation responsible for their safety had maintained the systems and infrastructure on which that safety depended. The commercial decisions that degraded those systems were made by people who were not indifferent to the welfare of the workforce, people who, if asked directly whether they were prepared to accept that their budget decisions might kill workers, would have answered no. They would not have recognised their decisions as decisions of that kind. They were making budget decisions, operating within the normal frameworks of corporate financial management. The deaths were an unintended consequence of those decisions. That does not make those decisions morally neutral.

At Wittenoom, the Production Gravity that shaped the company's management of the asbestos hazard resulted in the deaths of more than 2,000 people over a period of decades, deaths that occurred through diseases of excruciating difficulty and that were, from the point at which the medical evidence was available to the company, foreseeable. The commercial decisions that maintained the dust conditions at Wittenoom were made by people who knew, or who should have known, that those conditions were dangerous. The Production Gravity that shaped those decisions did not operate in the absence of knowledge. It operated in the presence of it.

The difference between Texas City and Wittenoom in this respect, between Production Gravity operating in the absence of full awareness of consequences and Production Gravity operating with at least partial awareness, is morally significant. It is a difference that the ethical analysis of Chapter 10 explores in depth. What both cases share, and what the Production Gravity concept captures, is the structural tendency of organisations under commercial pressure to make decisions that prioritise production over safety in ways that, in the aggregate and over time, produce conditions of genuine danger, whether or not the people making those decisions fully recognise what they are doing.

Chapter 8

Leadership Detachment

When Those at the Top Can No Longer See

How the structural distance between senior leaders and operational reality produces confident ignorance, and why that ignorance is the most dangerous condition of all: Chernobyl and Gretley.

The Paradox of the Informed Leader

Senior leaders in large organisations are, in a formal sense, better informed than almost anyone else in those organisations. They receive more reports, attend more briefings, review more performance data, and have access to more aggregated analysis of the organisation's activities than any single operational employee. This volume and scope of the information available to them are, by any conventional measure, impressive.

And yet, in disaster after disaster, the evidence reveals a consistent pattern: senior leaders whose confident belief in the safety and soundness of their organisation's operations was grounded not in accurate knowledge of the system's actual state, but in the picture that the system's reporting structures had constructed for them, a picture that was reassuring, professionally presented, and fundamentally disconnected from operational reality.

This is the paradox of Leadership Detachment. The more information flows upward through an organisation's hierarchy, the more opportunities there are for that information to be filtered, aggregated, reframed, and ultimately transformed from an accurate representation of the operational system's actual state into a version of that state that the hierarchy finds comfortable to present and comfortable to receive. Senior leaders who rely on those information flows, who assess their organisation's risk state by reference to the reports and metrics produced by the systems designed to inform them, may end up being the worst-informed people in the organisation about the things that matter most for safety.

Hopkins (2007) introduced the concept of mindful leadership to describe the alternative: leaders who actively and persistently probe beneath the surface of formal reporting systems, who seek out the operational reality of

DOI: 10.4324/9781042017539-11

their organisations rather than accepting the sanitised versions that those organisations' communication structures tend to produce. Mindful leaders are not distinguished primarily by their technical knowledge or their personal dedication to safety values, they are distinguished by their epistemic habits, by the ways in which they seek, test, and act on information about what is actually happening in the systems they are responsible for.

Leadership Detachment is the structural tendency toward the opposite of mindful leadership, the gradual, organisationally produced loss of contact between senior leaders and the operational reality they are nominally responsible for managing. This chapter examines that tendency through two case studies: the Chernobyl nuclear disaster of 1986, in which Leadership Detachment operated at both the plant level and the level of the Soviet state, and the Gretley mine disaster in New South Wales in 1996, in which a small management team's detachment from the technical realities of underground mining produced a catastrophic failure that was, in the most direct possible sense, the consequence of not knowing what any competent mine manager should have known.

Case Study One: Chernobyl, Ukrainian Soviet Socialist Republic, 26 April 1986

Case Overview: *At 1:23 am on 26 April 1986, Reactor No. 4 of the Chernobyl Nuclear Power Plant near Pripyat in the Ukrainian Soviet Socialist Republic exploded during a safety test, releasing a quantity of radioactive material approximately 400 times greater than that released by the atomic bomb dropped on Hiroshima. The immediate causes of the explosion were a combination of reactor design flaws and procedural violations during the test. The underlying causes included a culture of deference to authority that suppressed the upward communication of safety concerns, a leadership structure that prioritised production performance over safety management, and a political and institutional system that had, over decades, produced exactly the kind of confident ignorance at the senior leadership level that this chapter identifies as the defining characteristic of Leadership Detachment.*

The Soviet Nuclear Culture and Its Epistemic Consequences

Understanding Chernobyl requires understanding the institutional environment within which the Soviet nuclear power programme operated, an environment in which the normal mechanisms through which accurate information about operational reality might reach senior decision-makers had been systematically degraded by decades of political and institutional pressure.

Plokhy (2018) provides the most comprehensive historical account of the Chernobyl disaster and its institutional context available in English. His analysis makes clear that the conditions that produced the disaster were not primarily technical; the RBMK reactor design had known flaws, and those flaws contributed to the severity of the accident, but they do not explain why the operating procedures that might have managed those flaws were violated or why the safety culture of the plant permitted the violations to occur. The conditions that produced the disaster were institutional and cultural: the product of a system of governance and organisational management that had, over time, generated precisely the conditions for Leadership Detachment at every level from the plant floor to the Kremlin.

The Soviet institutional culture that shaped the Chernobyl plant's management was characterised, in its essential features, by a systematic bias toward the reporting of success and the suppression of the reporting of failure. In an institutional environment in which the acknowledgement of problems was career-limiting and the demonstration of production performance was career-enhancing, the information flows that should have carried accurate reports of operational reality upward through the organisation were shaped, at every level, by the incentive to present the most favourable possible picture of the system's state. This was not unique to the nuclear power industry. It was a pervasive feature of Soviet institutional culture that applied with particular force in sectors, like nuclear power, where the gap between official performance claims and operational reality was potentially most dangerous.

The consequence of this institutional culture for leadership at the Chernobyl plant was a management team whose confidence in the plant's operational performance was grounded in reports and metrics that reflected the officially approved picture of how the plant was operating rather than the actual state of its systems and culture. The plant's director, Viktor Bryukhanov, appears in the historical record as a man who genuinely believed, for the first hours after the explosion, that the damage to the reactor was manageable, a belief so far removed from the catastrophic reality that had already destroyed the reactor building that it can only be understood in the context of an institutional system that had produced, over years of consistent reinforcement, a leadership team genuinely incapable of receiving and processing information that contradicted its operationally optimistic worldview (Plokhy, 2018; Higginbotham, 2019).

The Test, the Shift, and the Authority Gradient

The proximate cause of the Chernobyl disaster was a safety test, an experiment designed to determine whether the turbines of the reactor could, in the event of a power outage, generate sufficient electricity to power the reactor's emergency cooling pumps during the brief period before the diesel backup generators reached operating speed. The test had been attempted on previous occasions and had not been successfully completed. On the night of 25 to 26 April 1986, it was attempted again, under the management of a shift led by Aleksandr Akimov and including the reactor control operators Leonid Toptunov and others.

The test required the reactor to operate at a specific, relatively low power level. During the preparation for the test, the reactor's power dropped far below the required level, to a point at which, under established operating procedures and in the judgement of qualified nuclear engineers, continuation of the test was unsafe. The reactor was in a condition known as xenon poisoning, in which the build-up of xenon-135 in the reactor core made the resumption of stable operation at the required power level technically problematic and potentially dangerous.

At this point, the established procedure was clear: the test should have been aborted and the reactor allowed to stabilise. The decision that was made instead, to continue the test despite the reactor's dangerously unstable condition, was made under pressure from Anatoly Dyatlov, the deputy chief engineer for operations, who had authority over the shift and who had been responsible for the test programme. Dyatlov's insistence on continuing the test, in the face of the concerns expressed by the shift operators, is one of the most documented examples in the disaster literature of what Dekker (2011) calls the authority gradient: the steep hierarchical differential between a senior manager and the operational personnel responsible for the physical system, which produces conditions in which the professional concerns of the operators are overridden by the institutional authority of the manager.

The authority gradient at Chernobyl was not simply a feature of Dyatlov's personal management style. It was a structural feature of the Soviet institutional culture within which the plant operated, a culture in which deference to authority was enforced through a combination of professional consequence and political pressure and in which the expression of safety concerns that contradicted a senior manager's operational judgement was a professionally risky act. The operators who expressed concern about the reactor's condition before the test were not simply deferring to Dyatlov's authority out of personal timidity. They were responding rationally to the institutional environment that shaped their professional options (Plokhy, 2018).

The State Level and Its Detachment

The Leadership Detachment at Chernobyl operated not only at the level of the plant's management but also at the level of the Soviet state. The response of the Soviet political and institutional leadership in the hours and days following the explosion – the initial denial of the severity of the accident, the delay in evacuating Pripyat, the deployment of workers and emergency responders without adequate protective equipment, and the suppression of accurate information about the radiation releases – was itself a manifestation of Leadership Detachment at the highest institutional level.

The Soviet leadership's initial response to the reports from Chernobyl was shaped by exactly the institutional epistemic culture that had produced the detachment within the plant. The reports coming up through the chain of command in the first hours after the explosion were filtered through the same bias toward reassurance that shaped all upward communication in the Soviet institutional system. Bryukhanov's initial reports, transmitted through the official channels to Moscow, significantly understated the severity of the accident, in part because Bryukhanov himself did not fully understand what had happened and in part because the institutional incentive structure made the reporting of catastrophic failure professionally and personally dangerous.

Gorbachev's later account of the Soviet leadership's response acknowledged that the institutional culture of the Soviet system had made it structurally incapable of receiving and acting on accurate information about the severity of the accident in time to minimise its consequences. The Leadership Detachment that had operated within the plant, the confident ignorance of a management team whose picture of the plant's operational state was grounded in the officially approved version of that state rather than its reality, was replicated at the level of the Soviet state, with consequences that extended far beyond the immediate disaster site (Plokhy, 2018).

Chernobyl and the Architecture of Confident Ignorance

Higginbotham's (2019) narrative account of the Chernobyl disaster and its aftermath provides a richly detailed picture of the human experience of Leadership Detachment, of the gap between what the leaders at every level of the Soviet system believed about the safety and performance of the Chernobyl plant and what the operational reality actually was. That gap was not produced by any single failure or any single decision. It was the accumulated product of decades of institutional pressure on the information flows that should have kept leadership connected to operational reality.

The concept of confident ignorance, a term that captures the specific quality of the Chernobyl leadership's detachment, in which the ignorance was not passive or uncertain but active and confident, is relevant beyond the Soviet context. It describes a condition that can develop in any organisation where the information flows from operational reality to senior leadership are systematically shaped by the incentive to reassure rather than to inform. The specific mechanisms of that shaping differ between the Soviet institutional system and contemporary commercial organisations. The essential dynamic is the same: leaders who believe they understand the risk state of their organisation because the information they receive tells them they understand it, when what the information actually reflects is not the operational reality but the picture of operational reality that the organisation's communication structures have found it appropriate or convenient to present.

> *The most dangerous leaders are not those who know their organisation is unsafe and choose to accept the risk. The most dangerous leaders are those who genuinely believe their organisation is safe, because the systems they rely on to tell them the truth have been systematically constructed to tell them what they want to hear.*

Case Study Two: Gretley Colliery, New South Wales, Australia, 14 November 1996

Case Overview: *On 14 November 1996, four miners were killed at the Gretley colliery near Newcastle in New South Wales when the panel they were mining broke through into an old, water-filled working. The old workings flooded the area in which the men were working, killing all four. The subsequent inquiry, and Hopkins' (2007) detailed analysis of the disaster, found that the breakthrough was the direct result of a failure by the mine's management to correctly identify the location of the old workings, a failure that should not have been possible given the information available to competent mine management, and that was, in Hopkins' analysis, the product of a specific form of Leadership Detachment: the occupation of senior mine management roles by people who lacked the technical competence required to understand the operational risks they were nominally responsible for managing.*

The Nature of the Hazard and the Available Information

Underground coal mining in the Hunter Valley and Newcastle coalfields of New South Wales had, by the 1990s, been conducted for more than a century. The result was an extensive network of old workings, abandoned mine tunnels, worked-out panels, and old roadways, many of which were flooded with groundwater and some of which contained methane and other gases. For any active mine operating in close proximity to these old workings, the accurate identification of their location was a fundamental safety requirement: a mine that broke through into old, water-filled workings without adequate preparation could, and occasionally did, flood the working area with catastrophic speed.

The tools available to mine management for identifying the location of old workings included historical mine plans, geological surveys, and the mathematical modelling of mining geometry. For the Gretley colliery, the relevant old workings, the Young Wallsend workings, abandoned decades earlier, were documented in historical mine plans that were, in principle, available to the mine's management. The mathematical relationship between the coordinates of the active Gretley workings and the recorded coordinates of the old Young Wallsend workings could have been calculated, and had that calculation been performed correctly, it would have revealed that the panel being mined on 14 November 1996 was approaching the old workings and that breakthrough was imminent.

The calculation was not performed correctly. The mine plans were consulted, but their use was inadequate. The coordinate systems in which the historical plans were recorded were not properly reconciled with the coordinate system being used for the active workings. The result was an error in the estimated position of the old workings that placed them significantly further from the active panel than they actually were, an error that gave the mine's management a false sense of security about the distance remaining before the panel would reach the danger zone (Hopkins, 2007).

Competence, Certification, and the Gap Between Them

Hopkins' (2007) analysis of the Gretley disaster is centrally concerned with a question that is, in the context of the Blindspot Model's fifth condition, directly relevant: what did the mine's management know, and what should they have known?

The mine's management, its under-manager and the persons responsible for the technical management of the mining operation held the statutory certifications required under New South Wales mining legislation to occupy their roles. The certification system was designed to ensure that people in positions of technical management responsibility in mines had demonstrated the competence required to manage the hazards of underground mining safely. In the case of Gretley, the system had produced a management team that held the required certifications but did not possess the practical technical competence, specifically, the ability to correctly use historical mine plans and coordinate systems to identify the location of hazardous old workings, that those certifications were supposed to represent.

Hopkins identifies this as a specific failure of the certification system: a gap between the formal credential and the actual competence it was supposed to certify. But he also identifies something more fundamental in the Gretley case, a form of Leadership Detachment that operated through the absence of technical knowledge rather than through the filtering of information. The managers at Gretley were not receiving accurate information and failing to act on it. They were receiving information, the historical mine plans and the surveying data that they lacked the technical competence to correctly interpret. Their detachment from the operational reality of their mine was not produced by the hierarchical filtering of upwards communication. It was produced by the absence, in the people occupying the relevant leadership roles, of the knowledge required to bridge the gap between the formal information available to them and the operational reality it described.

The Role of the Regulatory System

The Gretley disaster raises, as the Pike River disaster did, significant questions about the role of the regulatory system in producing or permitting the conditions that led to the disaster. The certification system that was supposed to ensure competent mine management had failed, not through regulatory negligence in the straightforward sense, but through a more systemic failure: the gradual devaluation of technical competence as the primary qualification for mine management, and its replacement by administrative and managerial qualifications that did not carry the same requirement for demonstrated technical expertise in the specific hazards of underground mining.

Hopkins (2007) traces this devaluation across the history of mine management certification in New South Wales, arguing that the regulatory system had, over time, shifted its emphasis from technical competence, demonstrated knowledge of the specific engineering and geological hazards of underground mining, towards administrative competence, demonstrated ability to manage systems, processes, and people. This shift was not without justification: the management of modern mine operations is genuinely complex and genuinely requires sophisticated administrative and managerial capability. But the substitution of administrative for technical competence at the regulatory level had produced a certification system that could no longer guarantee that the people certified to manage underground mines had the specific technical knowledge required to identify the hazards that killed the Gretley miners.

This regulatory dimension of Leadership Detachment is important for the argument of this chapter. The detachment of the Gretley mine management from the operational reality of their mine was not simply an individual failure or a corporate failure. It was, in part, a regulatory failure: the failure of the system designed to ensure that people in positions of mine management responsibility possessed the competence required to manage the mine's hazards. The Leadership Detachment that produced the Gretley disaster was, in this sense, institutionally produced, the consequence of regulatory choices that had, over time, redefined the competence requirements for mine management in ways that left the regulatory system unable to prevent the appointment of people who lacked the technical knowledge their roles required.

The Aftermath and Its Lessons

The coronial inquest and subsequent prosecutions following the Gretley disaster produced a range of findings and legal consequences for the individuals and organisations involved. The mine's under-manager and the surveyor responsible for the survey work were found to have been negligent. Fines were imposed. The regulatory system was reviewed and amended.

Hopkins' (2007) analysis provides a more penetrating account of what the disaster required in terms of organisational and regulatory learning. His central argument, that the Gretley case illustrates the concept of mindful leadership by negative example, demonstrating precisely what is absent when Leadership Detachment has occurred, is directly relevant to this chapter's argument. The mindful leader that Gretley needed was one who knew enough about the specific technical hazards of underground coal mining to ask the right questions about the proximity of old workings, to challenge the confidence of the survey results, and to understand that the consequences of an error in this domain were not recoverable. The leaders who were present did not have that knowledge. Their certification said they did. The four men who died in the flooded panel were the evidence that they did not.

The Structure of Leadership Detachment

Drawing on the Chernobyl and Gretley cases, and on the broader literature on leadership and organisational safety, the mechanisms through which Leadership Detachment develops and operates in organisations can be identified with some precision. Understanding these mechanisms is the foundation of the mindful leadership practice that represents the primary countermeasure to this condition.

Mechanism One: Hierarchical Information Filtering

The most pervasive mechanism of Leadership Detachment is the hierarchical filtering of information described in Chapter 4, the systematic transformation of safety-critical information as it moves upward through the layers of the organisation's reporting hierarchy. Each layer of the hierarchy aggregates and summarises the information it receives from below, presenting to the layer above a version of that information that is briefer, more orderly, and typically more reassuring than the raw operational data from which it was derived.

This filtering is not inherently malicious. It is the natural product of the need to manage information volume in large organisations. Senior leaders cannot attend to every incident report, every near-miss log, every maintenance record, and every safety concern raised at the operational level. Some degree of aggregation and filtering is not only inevitable but necessary. The problem arises when the filtering is so thorough, when the distance between the operational data and the senior leadership's reporting is so great, that the picture available to senior leaders bears little relationship to the operational reality from which it was derived. At that point, the leader's confidence in their organisation's safety performance becomes, in the most precise sense, unfounded. It is grounded not in knowledge but in the absence of information.

Mechanism Two: The Metric Substitution Problem

A specific and important form of hierarchical filtering is what can be called the metric substitution problem: the replacement of genuine understanding of the operational risk state with confidence in the metrics produced by the formal safety management system. Leaders who assess their organisation's safety performance by reference to formal metrics, lost time injury rates, near-miss reporting rates, audit completion rates, and leading indicator scores are assessing the performance of their safety management system, not the safety of their operations. These are related but not identical things, and the relationship between them depends critically on the quality, completeness, and honesty of the safety management system.

In organisations where the safety management system accurately captures and reflects the operational risk state, where the metrics are genuine measures of the things that matter for safety rather than measures of the organisation's compliance with its own reporting requirements, metric-based assessment of safety performance can provide useful, if incomplete, information about the organisation's risk state. In organisations where the safety management system has been shaped by the Production Gravity, cultural filtering, and Structural Secrecy described in earlier chapters, the metrics may bear very little relationship to the actual risk state. Leaders who rely on those metrics without interrogating their relationship to operational reality are leaders whose confidence in their organisation's safety performance is, in effect, a confidence in the performance of their formal systems rather than a confidence grounded in knowledge of the operational reality those systems are supposed to represent.

Weick and Sutcliffe (2001) argue that high-reliability organisations are distinguished, in part, by their leaders' resistance to the comfort of formal metrics, their insistence on testing the picture presented by formal systems against direct engagement with operational reality, and their institutional commitment to maintaining the sensitivity to operations that Weick calls one of the defining characteristics of mindful organisational management. This sensitivity is not a natural product of normal organisational processes. It requires specific cultural investment and specific leadership practice to maintain.

Mechanism Three: The Authority Gradient and Its Silencing Effect

The third mechanism, illustrated with particular clarity in the Chernobyl case, is the authority gradient: the steep hierarchical differential between senior leaders and operational personnel that, in cultures where deference to authority is strong, systematically silences the upward communication of operational concerns.

In organisations with a strong authority gradient, the concerns of operational personnel about the risk state of the system do not travel upward naturally. They are suppressed, sometimes by explicit discouragement, more often by the implicit understanding that raising concerns that contradict a senior leader's assessment is professionally risky. The result is a senior leadership team that receives a picture of the organisation's safety performance that has been systematically edited to remove the concerns, anomalies, and operational realities that the people closest to the work regard as most significant.

The authority gradient is not a Soviet institutional phenomenon. It operates, with varying degrees of intensity, in every organisation where hierarchical power differentials are strong and where the professional culture does not actively support the assertive upward communication of safety concerns. Edmondson's (2018) work on psychological safety, the shared belief that the environment is safe for interpersonal risk-taking, including the expression of concerns, questions, and dissenting views, is directly relevant here. Organisations with low psychological safety are organisations in which the authority gradient systematically suppresses the communication of operational reality to leadership. They are organisations in which Leadership Detachment is structurally produced by the culture of silence that the authority gradient generates.

Mechanism Four: Technical Competence Gaps

The fourth mechanism, the one most clearly illustrated by the Gretley case, is the technical competence gap: the absence, in the people occupying senior leadership roles, of the specific technical knowledge required to bridge the gap between the formal information available to them and the operational reality it describes.

This mechanism operates differently from the first three. The first three are mechanisms through which accurate information fails to reach leaders who would be capable of acting on it if it arrived. The fourth is a mechanism through which information does reach the leader, in some formal sense, but the leader lacks the knowledge required to correctly interpret its significance. The historical mine plans were available to the Gretley management. What was absent was the technical competence required to use them correctly. The information was present. The interpretive capacity was not.

Technical competence gaps in senior leadership are produced by the same set of forces that produce the certification-competence gap identified by Hopkins in the Gretley case: the progressive substitution of administrative and managerial qualifications for technical expertise in the definition of what senior management roles require. This substitution is, in many contexts, the product of entirely reasonable organisational choices and the recognition that senior leadership requires skills that go beyond technical expertise and that the most technically competent person in a function is not necessarily the best person to lead it. The risk is that those reasonable choices produce a leadership team that is administratively capable and technically detached: able to manage the formal systems that are supposed to represent the operational risk state, but unable to interrogate the gap between those systems and the operational reality they describe.

Voice, Trust, and the Social Architecture of Silence

The four mechanisms of structural detachment examined in the preceding section describe the architecture of the problem. They are conditions of information flow, organisational distance, signal filtering, and expertise. Taken together, they explain how senior leaders can come to hold a systematically inaccurate picture of the organisation they are responsible for. What they do not fully explain is why, in organisations where those structural conditions are partially mitigated by open-door policies, leadership visits, safety committees, and formal mechanisms for raising concerns, leaders so often still fail to learn what their people could tell them if they chose to.

The explanation is that Leadership Detachment is not only a structural phenomenon. It is also a social phenomenon. It is produced not only by the distance between the top and the bottom of the organisational hierarchy but also by the conditions – social, interpersonal, and psychological – that determine whether the people who hold safety-critical information will speak, whether they will be heard, and whether anything will happen if they do.

Morrison and Milliken (2000) identified this as organisational silence: the phenomenon by which employees in many organisations withhold information, concerns, and opinions about workplace problems in ways that are systematic, collectively reinforced, and typically invisible to leadership. Morrison and Milliken argued that organisational silence is not the simple aggregate of individual decisions to remain quiet. It is a collective phenomenon, sustained by shared beliefs about the risks of speaking up, the futility of being heard, and the social costs of being seen as disloyal, difficult, or a troublemaker. Those shared beliefs are learnt by watching what happens when others speak – and in organisations where speaking has historically produced adverse consequences for those who spoke, or no consequences at all for those whose warnings were ignored, silence becomes the rational, professionally protective, and socially reinforced default.

Argyris (1990) had identified a related dynamic some years earlier in his analysis of what organisations make undiscussable. His argument was that in many organisations, the most important problems – the ones that would require real examination, real candour, and real change – are precisely the problems that have become culturally undiscussable: they cannot be raised without violating implicit rules about what the organisation is willing to hear about itself. The undiscussability is not openly stated. It does not need to be. It is maintained by a thousand small signals – the careful reframing of difficult issues in meetings, the quiet sidelining of people who pressed too hard on uncomfortable questions, and the polite attention followed by no action – that teach the organisation's members where the conversational boundaries lie. Argyris' central insight was that organisations often become unable to solve their most serious problems not because the information required to solve them is unavailable, but because the conversational conditions required to discuss them openly have been foreclosed.

Edmondson's (2018) extensive work on psychological safety has direct bearing on this analysis. Psychological safety, in Edmondson's account, is the shared belief held by members of a team that the team is safe for interpersonal risk-taking: that one can raise a concern, admit an error, or challenge a prevailing view without being humiliated, punished, or professionally marginalised. Edmondson's research demonstrates that psychological safety is not a personality trait of individuals or a soft amenity of pleasant workplaces. It is a structural property of the social environment, produced or destroyed by the specific behaviours of leaders, and it is strongly predictive of whether teams will surface problems early or late. In the context of Leadership Detachment, psychological safety is the enabling condition without which the structural interventions described in the preceding section – the reporting systems, the site visits, the safety committees – will be consistently insufficient. A formal channel for raising concerns cannot do its work if the people who hold the concerns have concluded, on the basis of consistent social evidence, that raising them is professionally dangerous or pragmatically pointless.

Flin et al. (2008) provide an operational framework through which these social conditions can be understood in the specific context of high-hazard industries. Their analysis of non-technical skills – the cognitive, social, and personal-resource skills that underpin safe and effective performance in technically demanding environments – identifies three capacities that bear most directly on the conditions this chapter is concerned with: situation awareness, communication, and decision-making. Situation awareness, in their account, is not a matter of monitoring instruments. It is a matter of attending to the human and social fields as well as the technical field: reading the tone of a conversation, noticing that a team member has become unusually quiet, detecting that a briefing is being managed rather than disclosed. Communication, in their framework, is evaluated not only by what is said but by whether the social conditions for honest exchange have been established. Decision-making is understood as a collective process in which the quality of the outcome depends on the quality of the inputs received – which depends, in turn, on whether those inputs have been offered honestly. In each case, the skill at issue is a social skill, and each skill is assessable and trainable.

The application of this framework to senior leadership is direct. The skills that allow a frontline supervisor to read the state of a crew and know whether that crew is telling her the full picture are the same skills, operating at a different level, that allow a site leader to read the state of a supervisor's team and that allow a senior executive to read the state of a site. In each case, the skill is the capacity to attend to the interpersonal and psychological field as well as the technical field – to notice that the site manager is managing his briefing, that the tool-box talk is performing candour rather than enacting it, and that the crew is agreeing because agreement is what the visit is for.

The Deepwater Horizon record illustrates the consequences of the absence of these skills in the most concrete possible form. In the period before the blowout, senior executives of both BP and Transocean conducted a visit to the Deepwater Horizon platform. The visit included meetings with operational personnel and observation of operational activity, and the executives left the platform with a positive impression of its safety performance. The record of the investigations that followed (National Commission, 2011) indicates that, at the time of that visit, conditions on the platform were already materially degraded relative to the picture presented to the visitors: the pattern of operational pressure, compressed schedules, and ambiguous authority structures that would shortly produce the misinterpretation of the negative-pressure test was already well established. The visiting executives did not detect it. They did not detect it not because the evidence was unavailable in principle but because the social perception skills required to read the state of the platform – to notice the particular texture of compliance that is produced by pressure rather than confidence, to distinguish the briefing that is disclosing from the briefing that is managing – were absent. The visit, intended as a corrective to structural distance, instead compounded it, producing confident reassurance in place of the chronic unease that the developing conditions ought to have provoked.

The implication of this analysis is substantial. The countermeasure to Leadership Detachment that the next section describes – mindful leadership, in Hopkins' terms – cannot be produced by structural intervention alone. Channels of communication, no matter how well designed, will not carry honest information if the social conditions for honesty have not been established. Site visits, no matter how frequent, will not produce an accurate perception of operational reality if the visiting leaders do not possess the social perception skills to read what is in front of them. The structural work is necessary. It is not sufficient. The work of building an organisation in which accumulating danger can be seen is, in substantial part, the work of building the social conditions under which people will speak, and leaders will hear, and what is said and what is heard will be permitted to change what the organisation does.

Mindful Leadership as the Countermeasure

Hopkins' (2007) concept of mindful leadership, developed in direct response to the Gretley case but drawing on a broader theoretical foundation in Weick's work on sensemaking and the high-reliability organisations literature, provides the most developed account available in the applied safety literature of what the countermeasure to Leadership Detachment actually looks like in practice.

Mindful leadership, in Hopkins' account, is characterised by three defining practices. The first is active curiosity about the gap between systems as designed and systems as actually operating. Mindful leaders do not assume that the formal systems of their organisations, the procedures, the safety management systems, and the reporting structures are being implemented as intended. They probe actively for the gap between the formal picture and the operational reality, treating that gap as a permanent feature of complex organisations that requires continuous attention rather than a problem to be solved once and then set aside.

The second practice is direct engagement with operational reality. Mindful leaders do not rely exclusively on the formal reporting systems to tell them what is happening in their organisations. They seek direct contact with the operational environment through site visits, through conversations with frontline workers, and through the deliberate cultivation of communication channels that bypass the normal filtering mechanisms of the hierarchy. This direct engagement is not a substitute for formal reporting systems. It is a check on those systems, a means of testing whether the picture they produce is an accurate representation of the operational reality they are supposed to describe.

The third practice is what Hopkins calls a chronic unease: a persistent, professionally grounded unwillingness to accept that the absence of bad news is conclusive evidence of the absence of bad conditions. Mindful leaders understand that the information systems of their organisations are imperfect, that good news travels more reliably upward than bad news, and that the conditions for a serious incident can be developing in parts of the system that the formal reporting structures are not adequately capturing. Their chronic unease is not anxiety or dysfunction; it is the professional posture of someone who understands, from the evidence of the disaster literature, that confident belief in one's organisation's safety is itself a risk factor if that confidence is not continuously tested against operational reality.

Weick and Sutcliffe (2001) describe a related concept in their account of high-reliability organisations: what they call preoccupation with failure, the organisational habit of treating the absence of serious incidents not as evidence of safety but as evidence of the need for continued vigilance. In organisations characterised by preoccupation with failure, near-misses are not managed and set aside, they are treated as windows into the system's actual risk state, and the information they provide is actively pursued and acted on. This preoccupation is the organisational expression of the mindful leadership posture that Hopkins describes at the individual level.

Identifying Leadership Detachment in Your Organisation

The practical challenge of identifying Leadership Detachment is, as with Structural Secrecy, partly a function of its self-concealing nature. Leaders who are detached from operational reality are, by definition, not well positioned to recognise their own detachment. The indicators of Leadership Detachment must therefore be sought not primarily from the leaders themselves but from the gap between what leaders report believing about their organisation and what the operational evidence reveals.

The first indicator is the quality of leaders' specific knowledge of their organisation's current operational risk state. Leaders who can describe, with accuracy and specificity, the current major hazard risk profile of their operations, including the known deficiencies, the deferred maintenance items, the areas where operational practice deviates from procedure, and the concerns currently being raised by frontline workers, are leaders who have maintained meaningful contact with operational reality. Leaders who respond to these questions with reference to formal audit results, safety performance metrics, or general assurances about the quality of their safety management systems are leaders whose knowledge of operational reality is mediated entirely by formal reporting systems and whose detachment from the operational level is, accordingly, potentially significant.

The second indicator is the pattern of surprise when serious incidents reveal conditions that had been developing over an extended period. In organisations where Leadership Detachment is operating, senior leaders are typically genuinely surprised when a serious incident exposes conditions, deteriorated equipment, normalised procedural violations, and a safety culture markedly different from the one described in the organisation's formal communications that had been present in the operational environment for a significant time. The surprise is real, and it is the most direct possible evidence of the gap between the leadership's picture of the organisation and its operational reality.

The third indicator is the existence and quality of mechanisms for direct communication between senior leaders and frontline workers about safety concerns. In organisations where Leadership Detachment is being actively managed, where mindful leadership is being practised, there will be visible, functioning mechanisms for this communication: regular site visits by senior leaders who ask specific questions about safety conditions and listen carefully to the answers, structured processes for frontline workers to raise safety concerns directly with senior management, and cultural norms that make the expression of operational safety concerns to senior leadership a normal and professionally supported act. The absence of these mechanisms, or their presence in form without substance, is an indicator of Leadership Detachment.

The fourth indicator, and the one that requires the most organisational honesty to examine, is the history of the organisation's response to safety concerns raised through formal channels. In organisations where Leadership Detachment is operating through cultural filtering and authority gradient effects, concerns raised through formal channels will tend to be resolved at the level at which they were raised without escalation to senior leadership. The pattern of how the formal safety reporting system routes concerns, what gets escalated and what gets managed at the operational level is a direct map of the organisation's authority gradient and the degree to which that gradient is producing confident ignorance at the senior leadership level.

> *Leadership Detachment is not a character flaw. It is a structural condition produced by the ordinary functioning of large organisations. The leader who says, after a disaster, that they had no idea things were that bad is, in most cases, telling the truth. The question is not why they did not know. The question is why the organisation they were leading was designed in a way that made knowing the predictable outcome impossible.*

The End of Part II

The five chapters of Part II have examined each condition of the Blindspot Model in depth through detailed case study analysis. What the cases demonstrate, consistently and across a wide range of industries, countries, and eras, is the claim made in Chapter 3: that the five conditions are not independent risk factors but interlocking components of a system that, when all five are operating, produces a form of organisational blindness that is greater than the sum of its parts.

Signal Attenuation prevents the operational knowledge that exists at the frontline from reaching the decision-makers who could act on it. Normalised Tolerance revises the cultural standards against which that knowledge would be assessed even if it arrived. Structural Secrecy fragments the complete risk picture so that no individual or team ever possesses enough of it to recognise its full significance. Production Gravity erodes the safety margins that would otherwise provide protection against the consequences of the blindness. And Leadership Detachment produces, at the level of senior management, the confident ignorance that allows all four of the other conditions to operate without generating the corrective response they demand.

Part III examines what happens when all five conditions operate simultaneously and what the ethical obligations of the organisations and leaders who produce these conditions actually are.

Part III

When the Conditions Combine

Chapters 9 and 10

The perfect blindspot is not the sum of five failures. It is what five failures become when they stop being separate.

Part III examines what happens when the five conditions of the Blindspot Model operate not as discrete organisational risks but as a convergent system state. The integrated case analyses in Chapter 9 demonstrate that the most catastrophic organisational failures are not produced by any single condition in isolation but by the configuration that emerges when all five operate simultaneously, each amplifying and enabling the others.

Chapter 10 addresses the ethical dimension of this analysis. Systemic explanations of disaster carry a risk: they can become, in practice, mechanisms for distributing and thereby dissolving moral responsibility. Part III argues that the systemic account does not reduce individual and organisational accountability. It grounds it in a more accurate understanding of what organisations actually did, and failed to do, when they produced the conditions that made catastrophic harm predictable.

DOI: 10.4324/9781042017539-12

Chapter 9

The Perfect Blindspot

When All Five Conditions Operate Simultaneously

What happens when Signal Attenuation, Normalised Tolerance, Structural Secrecy, Production Gravity, and Leadership Detachment combine, Deepwater Horizon and the Herald of Free Enterprise.

When the Conditions Converge

The preceding five chapters have examined each condition of the Blindspot Model in isolation, tracing its theoretical foundations, demonstrating its operation through detailed case analysis, and identifying the specific mechanisms through which it develops and the characteristic indicators through which it can be detected. This analytical approach, while necessary for establishing the framework clearly, has an important limitation: it separates conditions that, in the most dangerous organisational states, do not operate separately. They operate together, each amplifying and enabling the others, in a configuration that produces a form of organisational blindness qualitatively more severe than any single condition can produce on its own.

This chapter examines what happens when all five conditions converge, when Signal Attenuation, Normalised Tolerance, Structural Secrecy, Production Gravity, and Leadership Detachment operate simultaneously within the same organisational system. The result is what this book calls the perfect blindspot: the organisational state in which no single condition is providing any compensating protection against the blindness produced by the others, in which the interactions between conditions are mutually reinforcing rather than mutually correcting, and in which the organisation's capacity to see the danger it is accumulating has been so completely degraded that catastrophic harm becomes not merely possible but, absent some external interruption, essentially inevitable.

Two cases are examined in this chapter in which all five conditions can be clearly identified operating simultaneously and in interaction: the Deepwater Horizon disaster, revisited here in integrated analytical depth after its partial treatment in Chapter 4, and the capsize of the Herald of Free Enterprise

DOI: 10.4324/9781042017539-13

off Zeebrugge in 1987. These cases are chosen not because they are the only examples of the perfect blindspot in the disaster literature; Chapter 10 will identify moral and ethical dimensions visible across multiple cases, but because they are among the cases in which the documentary record is sufficiently detailed to support the kind of comprehensive, integrated analysis that demonstrating all five conditions simultaneously requires.

The purpose of this analysis is not simply descriptive. It is demonstrative: to show that the Blindspot Model, applied in integrated form to real cases, produces an account of why those disasters occurred that is more complete, more analytically coherent, and more practically useful than any single-condition explanation can provide. And to show that the disasters were not, as they are sometimes characterised, extreme outliers produced by exceptional circumstances. They were the predictable outcomes of identifiable organisational states, states that the Blindspot Model can identify and that active foresight can interrupt.

Case Study One: Deepwater Horizon, The Integrated Analysis

Case Overview: *The Deepwater Horizon disaster of 20 April 2010 has been examined in this book through the lens of Signal Attenuation in Chapter 4 and through the lens of Production Gravity in Chapter 7. This chapter examines it through all five lenses simultaneously, demonstrating how the five conditions of the Blindspot Model interacted within BP's deepwater drilling operations to produce a perfect blindspot, an organisational state in which the warnings that were present within the system could not be assembled, transmitted, or acted upon in any form that would have interrupted the trajectory toward catastrophe.*

Signal Attenuation at Macondo: The Full Picture

The Signal Attenuation at Macondo, examined in detail in Chapter 4, was not confined to the final hours before the disaster. It operated across the entire lifecycle of the well, from the design decisions made months before the blowout to the real-time reclassification of the anomalous negative pressure test result on the afternoon of 20 April.

At the design stage, engineering concerns about the well's casing design, specifically, about the number of centralisers used and the implications for the quality of the cement job, were raised and not adequately acted upon. The signal was attenuated at the project management level: received, acknowledged, and effectively overridden by a decision that prioritised the schedule and cost implications of the concern over its safety substance. At the operational stage, concerns about the well's pressure behaviour during testing were raised and reclassified through the bladder effect rationalisation described in Chapter 4. At every stage, the information that would have supported a more cautious decision existed within the system and failed to reach the decision-making level in the form required to produce the caution it warranted.

The National Commission (2011) found that the Signal Attenuation at Macondo was not the product of a single communication failure or a single management decision. It was the product of communication structures across BP's deepwater drilling operations that systematically privileged operational momentum over the honest upward transmission of safety concerns. Those structures were themselves the product of a corporate culture, documented by Hopkins (2012) and the National Commission in considerable detail, in which the professional identity of the people involved in deepwater drilling operations was closely bound up with the ability to manage complex wells efficiently and complete them on schedule. In that culture, the escalation of safety concerns was, in practice, a professionally costly act, one that required the individual raising the concern to challenge the operational momentum of a multimillion-dollar project and to accept the professional consequences of doing so.

Normalised Tolerance at Macondo: The History of Accepted Risk

The Normalised Tolerance that operated in the context of the Macondo well was both specific to that well and embedded in the broader culture of deepwater drilling operations as BP practised them. At the well-specific level, the series of complications that had occurred during the Macondo drilling, the lost returns, the stuck pipe, and the unexpected gas influx had each been managed without catastrophic consequence and had each, through the ordinary process of operational experience, generated a revised understanding of what the well could tolerate. Each managed complication became, in the interpretive culture of the drilling team, evidence of competence rather than evidence of accumulating hazard.

At the broader cultural level, the practices that had developed across BP's deepwater drilling operations – the willingness to accept fewer barriers than the recommended standard, the tendency to rely on single barriers where multiple were specified, and the acceptance of cement jobs without full verification of their integrity – had become, over years of operational practice, the way deepwater drilling was done in BP's Gulf of Mexico operations. These practices had been applied across multiple wells without producing blowouts. Their repeated application without catastrophic consequence had normalised them as acceptable operational standards. The deviance from the recommended barriers and verification standards had been absorbed into the organisation's understanding of what competent deepwater drilling looked like (National Commission, 2011).

Vaughan's (1996) normalisation of deviance framework maps directly onto this pattern. The production decisions that shaped the Macondo well's final barrier configuration were not made by people who believed they were accepting unacceptable risk. They were made by people who had learned, through years of operational experience in a culture shaped by Production Gravity, that the practices they were applying were within the range of what their organisation regarded as acceptable. The normalisation had done its work. The deviance was invisible as deviance.

Structural Secrecy at Macondo: The Fragmented Risk Picture

The Structural Secrecy that operated in the Macondo context was, as the National Commission documented in detail, a function both of the fragmentation of safety-critical information across the multiple organisations involved in the operation – BP as the lease operator, Transocean as the rig contractor, and Halliburton as the cementing contractor – and of the fragmentation of that information within BP's own organisational structure.

The well design decisions were made by BP's drilling engineering team in Houston. The operational decisions were made by the BP company men and Transocean well site leaders on the platform. The cementing decisions were made by Halliburton's engineers, working within a contractual relationship with BP that did not provide clear lines of responsibility for the integrated management of well integrity. The regulatory oversight was provided by the Minerals Management Service, which had, as the National Commission found, developed a relationship with the industry it regulated that had compromised its capacity to provide independent safety assurance. Across this network of organisations and relationships, the complete picture of the Macondo well's risk state, the accumulation of design decisions, operational complications, barrier inadequacies, and testing anomalies that together constituted the conditions for the blowout were held in fragments by different parties, none of whom had the complete picture, and none of whom had the structural position or the institutional incentive to assemble it (National Commission, 2011).

Within BP's own organisation, the Structural Secrecy was reinforced by the communication gap between the operational level, the platform and the project management team, and the senior leadership of BP's Gulf of Mexico operations. The National Commission found that senior BP executives had limited awareness of the specific well design and abandonment decisions that contributed to the disaster. This was not because the information was being concealed from them. It was because the organisational structures of BP's deepwater drilling operations did not routinely transmit that level of operational detail to the senior leadership and because the senior leadership did not have in place the direct engagement mechanisms that would have compensated for what those structures excluded.

Production Gravity at Macondo: The Commercial Context

The Production Gravity operating at Macondo has been documented in Chapter 7 in the context of BP's broader cost-reduction culture, but its specific manifestation in the Macondo well deserves integrated treatment here because it shaped, directly and concretely, each of the decisions in the causal sequence that led to the blowout.

The well was, at the time of the disaster, approximately 43 days behind schedule and approximately 58 million dollars over budget. Each day the Deepwater Horizon remained on the Macondo location was costing BP approximately one million dollars in rig rental. The commercial pressure to complete the temporary abandonment and release the rig was, by every account of the events reviewed by the National Commission, a significant and explicitly acknowledged feature of the decision-making environment. It did not produce a decision to take an unacceptable risk, in the sense that no one involved would have described their decision in those terms. It produced a series of decisions, each assessed within the context of the prevailing Normalised Tolerance and the attenuated information environment, each of which individually appeared defensible and which collectively produced a well that was catastrophically underprotected.

The Production Gravity at Macondo was not operating in isolation. It was operating through the Normalised Tolerance that had revised the organisation's understanding of what constituted adequate barrier protection. It was operating through the Signal Attenuation that prevented the engineering concerns about that barrier protection from reaching the decision-makers with sufficient weight to interrupt the commercial momentum. It was operating through the Structural Secrecy that prevented any single person from assembling the complete picture of the barrier inadequacies that the production-gravity decisions had produced. And it was operating through the Leadership Detachment that left BP's senior management without the operational knowledge required to recognise that the decisions being made on the Macondo well were not within the range of what their organisation's safety standards actually permitted.

Leadership Detachment at Macondo: The Invisible Accumulation

The Leadership Detachment that completed the perfect blindspot at Macondo operated at multiple levels. At the platform level, the BP company men and Transocean wellsite leaders who were the most senior decision-makers present had been detached, by the combination of production pressure and Normalised Tolerance, from the accurate understanding of the well's risk state that would have been required to make different decisions. They had the information, the anomalous pressure test result, and the history of barrier decisions, but they were operating within an interpretive framework shaped by the other four conditions that prevented them from seeing that information as the warning it was.

At the corporate level, the detachment was more structural. BP's senior Gulf of Mexico leadership did not have, and had not built, the direct operational engagement mechanisms that would have provided them with an accurate picture of the risk management practices being applied across their deepwater drilling operations. The formal reporting systems provided aggregate performance data. The direct operational knowledge of specific well design and abandonment practices, the knowledge that would have been required to identify the Normalised Tolerance and assess its implications, was not reaching the senior level. The leadership was detached, and the detachment was structural rather than personal (Hopkins, 2012).

The interaction of all five conditions at Macondo produced, in Reason's (1997) terms, the alignment of the Swiss cheese holes that allowed the trajectory of accident opportunity to pass through every layer of defence simultaneously. But unlike Reason's metaphor, which leaves the origin of the holes unspecified, the Blindspot Model identifies exactly where those holes came from. Signal Attenuation removed the warning signals from the decision-making level. Normalised Tolerance revised the standards against which the risks were assessed. Structural Secrecy prevented the complete picture from ever being assembled. Production Gravity created the decisions that moved the organisation toward its limits. Leadership Detachment removed the capacity to see that any of this was occurring. The holes did not appear randomly. They were produced, systematically, by the five conditions operating in combination.

Signal Attenuation: Engineering concerns were softened and overridden at the project management level; negative pressure test anomalies were reclassified in real time; and upward communication of safety concerns was structurally discouraged by professional culture.

Normalised Tolerance: Barrier protection practices normalised below recommended standards across multiple wells; cement job verification accepted without confirmation of integrity; each managed complication reinterpreted as evidence of competence.

Structural Secrecy: Risk picture was fragmented across BP, Transocean, Halliburton, and the regulator; no single party held the complete barrier picture; senior leadership was structurally excluded from operational detail.

Production Gravity: 43 days behind schedule; one million dollars per day rig cost; each barrier decision shaped by commercial pressure to complete and release the rig.

Leadership Detachment: Platform decision-makers operating within a normalised interpretive framework; corporate leadership without operational knowledge of specific best practices; no direct engagement mechanisms compensating for hierarchy filter.

At Macondo, all five conditions were operating simultaneously and in interaction. No single condition, addressed in isolation, would have been sufficient to prevent the disaster. Addressing any one of them would have required confronting the others. That is what the perfect blindspot means in practice: a state of organisational blindness so complete that no partial intervention is adequate; what is required is the comprehensive, integrated approach to active foresight that this book's final section describes.

Case Study Two: Herald of Free Enterprise, Off Zeebrugge, 6 March 1987

Case Overview: *On the evening of 6 March 1987, the roll-on roll-off passenger ferry Herald of Free Enterprise left the Belgian port of Zeebrugge with its bow doors open and capsized approximately 90 seconds later, killing 193 passengers and crew. The immediate cause was the failure to close the bow doors before departure. The organisational causes, examined in the Sheen Inquiry and analysed extensively in the subsequent safety literature, demonstrate the operation of all five conditions of the Blindspot Model within a passenger ferry operating company that had, over years of normal commercial operation, accumulated the conditions for catastrophic harm without anyone at any level of the organisation being able to see, from where they stood, the complete picture of what was developing.*

The Townsend Thoresen Culture and Its Five Conditions

Townsend Thoresen, the operating company of the Herald of Free Enterprise, was not an organisation with an obvious history of safety failure. Its ferries had been operating cross-channel routes for years. Its safety record, measured by the conventional metrics of maritime operations, was acceptable by industry standards. The Herald of Free Enterprise was not a poorly maintained vessel operated by incompetent crew. It was a vessel that had been operating in a commercial environment that had, over years of normal operation, produced the five conditions of the Blindspot Model in combination and had done so without generating the kind of visible warning signals that would have attracted serious external scrutiny or prompted internal alarm.

The Sheen Inquiry, which examined the disaster in considerable depth, produced the finding, quoted in Chapter 2 of this book, that the capsizing was the result of what the inquiry memorably described as a disease of sloppiness at every level of the corporation from the boardroom downwards. That finding is accurate as far as it goes, but it does not fully capture the systemic character of the conditions the inquiry documented. The disease was not simply sloppiness. It was the specific combination of organisational conditions, the five conditions, that this book has been describing across the preceding chapters.

Signal Attenuation: The Open-Door Reports

The possibility of sailing with bow doors open, the precise scenario that caused the disaster, had been raised as a safety concern within Townsend Thoresen before the Herald capsized. There were recorded instances of vessels sailing with bow or stern doors open or partially open prior to 6 March 1987. These instances had been observed; reported through whatever informal channels existed and had not produced the systematic response, a formal reporting requirement, a verification procedure, or an indicator system in the wheelhouse that would have been required to prevent a recurrence.

More specifically, a suggestion had been made within the company, before the disaster, that an indicator light should be installed in the wheelhouse to show the officer of the watch whether the bow doors were closed. This suggestion, which would have directly prevented the specific mechanism of the Herald disaster, had been raised through the company's internal communication channels and had not been actioned. The Sheen Inquiry found evidence in the company's documentation of this suggestion and its non-response. The signal, the recognition that the wheelhouse lacked positive confirmation of door closure status, had been transmitted. It had been attenuated before it could produce the corrective action it identified (Sheen, 1987).

The attenuation of this signal was not the product of a single management failure. It was the product of a communication culture within Townsend Thoresen that had, as the Sheen Inquiry found, allowed operational safety concerns to be raised and not answered without generating the escalation or accountability that those concerns required. The culture of not formally responding to operational safety communications, of treating them as informal suggestions rather than as safety-critical inputs requiring substantive response, was itself a form of systemic Signal Attenuation embedded in the organisation's operational practice.

Normalised Tolerance: The Accepted Risk of Departure

The departure of the Herald of Free Enterprise with its bow doors open was not, as noted above, without precedent. Previous instances of vessels departing with doors open or incompletely closed had occurred within the fleet. None had produced catastrophic harm, not because the risk was not real, but because the conditions that would make it catastrophic had not been present on those occasions.

Each previous incident without catastrophic outcome had, through the normalisation mechanism, contributed to a revised understanding of the seriousness of the open-door risk. The formal procedures required the doors to be closed before departure. But the cultural understanding of the departure process, the shared operational norms that shaped how the crew managed the pre-departure sequence under the commercial pressure of rapid turnaround schedules, had been revised, through repeated exposure to departures that were not quite within the procedural specification but that had not produced disaster, to accommodate a degree of procedural latitude that the formal system did not permit.

This Normalised Tolerance was not visible as Normalised Tolerance from inside the organisation. It was visible as the way experienced ferry crews managed the practical demands of rapid commercial operations as professional competence in managing the gap between ideal procedural compliance and operational reality. The deviance was invisible as deviance because it had been embedded, through years of operational practice and organisational accommodation, in the professional culture of the fleet.

Structural Secrecy: The Absent Verification System

The Structural Secrecy at the Herald of Free Enterprise operated through a specific and, in retrospect, straightforward mechanism: the absence of any positive confirmation system that would have allowed the officer of the watch on the bridge to verify that the bow doors were closed before ordering the vessel's departure.

The responsibility for closing the bow doors rested with the assistant bosun. The responsibility for verifying that they had been closed rested with the officer responsible for the car deck. The officer responsible for giving the order to depart rested with the officer of the watch on the bridge. In the chain of responsibility that ran from the assistant bosun to the bridge, there was no positive confirmation system, no indicator light, no formal sign-off procedure and no redundant verification step that would have ensured that the knowledge held by the person responsible for closing the doors was transmitted to the person responsible for authorising departure.

On the night of 6 March, the assistant bosun fell asleep before closing the doors. The officer responsible for verifying closure left the car deck to go to his departure station, having not verified that the doors were closed. The officer of the watch on the bridge ordered departure without positive confirmation that the doors were closed, because the ship's system of responsibilities provided no mechanism for obtaining such confirmation. The knowledge that the doors were open, which was, in principle, available to anyone who looked at the bow of the vessel, was not transmitted to the bridge because the Structural Secrecy of the vessel's departure verification system made that transmission dependent on a chain of individual actions, any one of which, if not performed, broke the chain entirely (Sheen, 1987).

The Structural Secrecy here was architectural: built into the physical and procedural design of the vessel's departure system. It was not the product of organisational complexity or geographic fragmentation. It was the product of a system design that assumed every step in the verification chain would be performed, without providing any structural redundancy to detect and compensate for the failure of any single step. That assumption was the organisational blindspot. The suggestion of an indicator light, which had been made and not actioned before the disaster, would have addressed it directly. The Signal Attenuation that prevented that suggestion from being implemented allowed the Structural Secrecy to persist until it became fatal.

Production Gravity: The Commercial Pressure of Cross-Channel Ferries

The commercial environment of cross-channel ferry operations in the late 1980s was intensely competitive. Townsend Thoresen and its competitors were operating on high-frequency schedules, with turnaround times in port that were under constant commercial pressure to be minimised. The time spent in port, loading, unloading, and preparing for departure, was time not spent carrying paying passengers, and the competitive economics of the route created a powerful gravitational pull toward the minimisation of turnaround time at every step of the departure process.

The Production Gravity in this context operated on the pre-departure procedures in a specific and identifiable way. The formal procedures required a specific sequence of checks before departure. The commercial pressure to minimise turnaround time created an incentive to perform those checks as rapidly as possible, and, where the operational culture permitted, to abbreviate or omit steps that experience suggested were, in practice, redundant. The closing and verification of the bow doors was not, in the operational culture of Townsend Thoresen, a step that carried the same weight as the formal procedures specified it should carry. The Normalised Tolerance had revised its salience. The Production Gravity had created the incentive for its abbreviation. The Structural Secrecy meant that the failure to perform it went undetected until it was too late.

Dekker (2011) has analysed the Herald case as a clear example of drift into failure: the gradual erosion of procedural compliance through the accumulation of individually reasonable accommodations to commercial pressure, each of which moved the operational norms of the departure process progressively further from the formal specification, until the gap between the formal requirement and the operational reality was wide enough to allow the specific failure that caused the disaster. The Production Gravity at Townsend Thoresen was not the dramatic, high-stakes commercial pressure of a company in financial crisis. It was the ordinary, continuous pressure of a commercial operation competing on schedule efficiency, the kind of pressure that operates in every commercial organisation and that, in the absence of effective countermeasures, produces exactly the procedural erosion that the Herald disaster illustrates.

Leadership Detachment: The Boardroom and the Bow Doors

The Sheen Inquiry's finding of a disease of sloppiness from the boardroom downwards was a finding of Leadership Detachment in all but name. The senior leadership of Townsend Thoresen was not operationally engaged with the safety management of its fleet in the way that the Blindspot Model's analysis of Leadership Detachment would require. The formal safety management systems that existed within the company did not provide the senior leadership with an accurate picture of the operational safety culture of its vessels. The direct engagement mechanisms that would have compensated for the hierarchy filter, the senior management visits to vessels, the direct conversations with operational crew, and the systematic probing of the gap between formal procedures and operational practice were not features of the company's leadership culture.

The suggestion of the indicator light, raised from the operational level and not actioned, is perhaps the clearest single piece of evidence of the Leadership Detachment at Townsend Thoresen. A leadership team that was genuinely engaged with the operational safety management of its fleet would have had a mechanism through which a suggestion of that kind, identifying a specific gap in the vessel's pre-departure verification system, would have been received, evaluated, and either implemented or rejected on clear safety grounds. The absence of such a mechanism and the non-response to the suggestion indicate a leadership culture in which operational safety concerns were not receiving the senior attention that the management of a major hazard passenger vessel operation demands.

The Sheen Inquiry documented other evidence of Leadership Detachment at the senior level. The company's response to previous open-door incidents, incidents that should have generated a systematic review of pre-departure verification procedures, had not produced the kind of mindful leadership response that Hopkins (2007) describes: the active probing beneath the incident to identify the systemic conditions that had produced it and the systemic changes required to prevent its recurrence. Each incident had been managed at the operational level. Each had generated a brief response and no lasting structural change. The Leadership Detachment that prevented those incidents from producing the systemic response they warranted was the same detachment that, on 6 March 1987, left the organisation without any structural protection against the specific failure that killed 193 people.

Signal Attenuation: Indicator light suggestion was raised and not actioned; open-door incidents were managed at the operational level without systematic escalation; and the communication culture did not require a formal *response to safety concerns.*

Normalised Tolerance: Previous open-door departures without catastrophic outcomes revised operational understanding of procedural compliance requirements; the departure sequence was culturally abbreviated under turnaround pressure.

Structural Secrecy: No positive confirmation system between car deck and bridge; departure verification dependent on an unbroken chain of individual actions with no structural redundancy; bow door status unknown to officer of the watch.

Production Gravity: Intense commercial pressure to minimise turn-around time; competitive cross-channel environment; pre-departure procedure compliance eroded by schedule pressure across the fleet.

Leadership Detachment: Senior leadership is not operationally engaged with fleet safety culture; there is no mechanism for safety suggestions to receive systematic senior review; and previous incidents are not producing a mindful systemic response.

What the Perfect Blindspot Teaches

The integrated analysis of Deepwater Horizon and the Herald of Free Enterprise through all five conditions of the Blindspot Model produces several findings that are important for the argument of this book and for the practical framework developed in its final part.

Finding One: The Conditions Are Not Additive; They Are Multiplicative

When all five conditions are operating simultaneously, the resulting organisational blindness is not simply the sum of the five individual conditions. It is qualitatively different, a compound state in which each condition amplifies the effects of the others and in which the normal corrective mechanisms that might interrupt any single condition are themselves degraded by the presence of the others.

Signal Attenuation prevents the warning signals that might interrupt Normalised Tolerance from reaching the decision-making level. Normalised Tolerance provides the interpretive framework within which attenuated signals, if they do arrive, are assessed as acceptable. Structural Secrecy prevents the complete picture that might override both attenuation and normalisation from being assembled. Production Gravity provides the motivational force that drives the organisation toward its limits despite all three. And Leadership Detachment removes the epistemic capacity that is the ultimate safeguard against all four. The five conditions together produce a system without any functioning corrective mechanism, a blindspot that is, in the most literal sense, total.

Finding Two: The Perfect Blindspot Is Detectable Before the Disaster

Both case studies demonstrate that the five conditions were operating and detectable, in principle, before the disasters that revealed them. At Macondo, the engineering concerns, the normalised barrier practices, the fragmented risk picture, the commercial schedule pressure, and the absence of senior operational engagement were all features of the organisational landscape that a systematic assessment using the Blindspot Model would have identified before 20 April 2010. At the Herald, the previous open-door incidents, the unanswered indicator light suggestion, the production pressure on turnaround times, and the absence of senior safety engagement were all observable before 6 March 1987.

The disasters were not invisible in advance. They were unseen, which is a different thing. The information required to see them was present. What was absent was the analytical framework for assembling it into the complete picture and the organisational capacity to act on that picture if it had been assembled. The Blindspot Model provides the framework. The active foresight approach developed in Chapter 12 provides the organisational capacity. Together, they constitute the tools required to see what, in both these cases, went unseen until it was too late.

Finding Three: Addressing the Perfect Blindspot Requires Integration

Perhaps the most important practical finding from the integrated analysis of the perfect blindspot is that addressing it requires an integrated response. Addressing Signal Attenuation alone, improving upward communication channels without addressing the Normalised Tolerance that shapes how the signals that arrive are interpreted, will not interrupt the blindspot. Addressing Normalised Tolerance alone, resetting the standards of acceptable practice without addressing the Structural Secrecy that prevents the complete picture of deviation from those standards from being visible, will not interrupt it. Addressing any single condition in isolation will reduce the severity of the blindspot at that specific point while leaving the other four conditions in place to compensate for the reduction.

This finding has profound implications for the way organisations design their safety improvement programmes. Most organisational responses to disaster, and most safety improvement initiatives that are not responses to disaster, address specific conditions: they improve a communication system, or reset a safety standard, or restructure a safety function, or commission a safety audit. These are valuable interventions. They are insufficient as stand-alone responses to the perfect blindspot because they do not address the system of mutually reinforcing conditions that produces the blindspot's most dangerous form.

The active foresight framework developed in Chapter 12 is designed to provide an integrated response to all five conditions simultaneously, not because they must always be addressed together, but because the evidence from the perfect blindspot cases demonstrates that addressing them in isolation is not adequate when all five are operating. Partial solutions to a total blindspot produce a safer organisation at the specific points they address. They do not produce an organisation that can see.

> *The perfect blindspot is not an extreme outlier. It is the predictable outcome of five ordinary organisational conditions operating in combination. Deepwater Horizon and the Herald of Free Enterprise were not exceptional organisations producing exceptional failures. They were ordinary organisations, capable, and commercially successful, staffed by competent professionals, that had accumulated, through entirely ordinary organisational processes, the conditions that made catastrophe not just possible but, in the absence of intervention, essentially certain.*

The Predictability Argument

The most important claim that the integrated analysis of the perfect blindspot supports, and the claim on which the practical value of this book depends, is the predictability argument: the assertion that the conditions examined in this chapter were identifiable before the disasters they produced and that an organisation equipped with the Blindspot Model and the active foresight framework could, in principle, have identified and interrupted them in time to prevent catastrophic harm.

This is a strong claim, and it deserves to be stated carefully. The predictability argument does not assert that the specific disaster, the specific event, on the specific date, in the specific form it took, was predictable. Complex systems fail in complex ways, and the specific mechanisms of individual disasters are rarely foreseeable in detail. What was predictable, and what the Blindspot Model claims can be identified in advance, was not the specific disaster but the organisational state that made a disaster of some kind not merely possible but, over time, likely.

Both Macondo and the Herald were organisations in which, for extended periods before the disasters that revealed them, the five conditions of the Blindspot Model were operating. Both were organisations in which, had someone been looking with the right analytical framework, the signature of the perfect blindspot was visible in the organisational record. Both were organisations in which the active foresight that Chapter 12 describes would have identified the conditions and created the opportunity – not the certainty, but the opportunity – for intervention before the precipitating events that triggered the disasters.

That opportunity is what this book is about. Not the certainty of prevention, complex systems will always retain some irreducible capacity for failure. But the genuine, evidence-based, practically achievable possibility of seeing the danger that is developing in time to address it if the organisation has the frameworks, the habits, and the institutional will to look.

Chapter 10

Moral Drift and Ethical Corrosion

How organisations arrive at decisions that cause catastrophic harm without intending to, and why that does not exempt them from moral accountability.

The Problem of Unintended Harm

One of the most persistent tensions in the analysis of organisational disaster is the tension between systemic explanation and moral accountability. The systemic account, the account this book has been developing across the preceding nine chapters, argues that the conditions producing catastrophic harm are organisationally generated, that they develop through ordinary organisational processes rather than through exceptional individual failures, and that understanding them requires a framework focused on systems and conditions rather than on individual actors and their decisions. This account is well supported by the evidence. It is also, for some readers, morally uncomfortable.

The discomfort is understandable. If the conditions for disaster are organisationally produced, if they are the predictable outputs of structures, cultures, and pressures that operate through the ordinary functioning of complex enterprises, then where does individual moral accountability sit? If the managers who made the decisions that produced the conditions for Macondo, or for Texas City, or for Wittenoom, were operating within organisational contexts that made those decisions locally rational, does the systemic account reduce their moral responsibility for the harm their decisions caused? Does understanding why the Blindspot developed somehow excuse those who allowed it to develop?

The answer this chapter gives is an unambiguous no. The systemic account does not reduce individual moral accountability. What it does is relocate it, from the specific individuals who made the proximate decisions to the broader set of individuals who, at every level of the organisations examined in this book, made the decisions that produced the five conditions. Those decisions were moral choices. The fact that they were made within organisational

DOI: 10.4324/9781042017539-14

contexts that made them locally rational does not strip them of their moral character. It contextualises that moral character; it helps us understand how morally committed people arrived at morally harmful outcomes, but it does not dissolve it.

This chapter examines the ethical dimension of the Blindspot Model directly. It introduces the concept of moral drift, the gradual erosion of ethical standards and moral attention that occurs through the same incremental, self-reinforcing mechanisms as the five conditions of the Blindspot Model, and argues that moral drift and the Blindspot are not separate phenomena. They are the same phenomenon examined from different analytical angles. The Blindspot is what organisational moral drift looks like from the outside, through the lens of safety science. Moral drift is what the Blindspot looks like from the inside, through the lens of ethics. Together, they produce a more complete account of how organisations harm people without intending to and of what that means for the moral responsibilities of the people who lead them.

Moral Drift: The Ethical Parallel to Normalised Tolerance

The concept of moral drift is most clearly understood in parallel with Normalised Tolerance, the second condition of the Blindspot Model. Just as Normalised Tolerance describes the gradual revision of safety standards through repeated exposure to risk without immediate catastrophic consequence, moral drift describes the gradual revision of ethical standards through repeated exposure to morally compromised decisions without immediate ethical accountability.

Bandura (1999) developed the concept of moral disengagement to describe the psychological and social mechanisms through which individuals and organisations progressively disengage from the moral standards that would normally govern their behaviour. His framework identifies eight specific mechanisms of moral disengagement, moral justification, euphemistic labelling, advantageous comparison, displacement of responsibility, diffusion of responsibility, dehumanisation, attribution of blame, and distortion of consequences, each of which represents a specific way of reframing morally problematic conduct so that it no longer triggers the ethical attention and the behavioural restraint that the moral standard would normally produce.

In the organisational disaster context, these mechanisms of moral disengagement operate through the same structural channels as the five conditions of the Blindspot Model. The Production Gravity that drives cost-reduction decisions provides the moral justification: the organisation must reduce costs to survive, to protect shareholder value, to maintain employment, and to serve its broader economic purpose. The Signal Attenuation that prevents the human consequences of those decisions from reaching the decision-making level removes the information that would be required to trigger the distortion of consequences mechanism. The Normalised Tolerance that embeds deviant practices in the organisational culture provides the advantageous comparison: everyone in the industry operates this way; we are no worse than our competitors. The Structural Secrecy that fragments the risk picture diffuses the responsibility: no single person made the decision that produced the harm, so no single person is responsible for it. The Leadership Detachment that leaves senior leaders without accurate knowledge of the operational reality displaces the responsibility downward: the people on the ground made the operational decisions; they bear the operational accountability.

Tenbrunsel and Messick (2004) introduced the concept of ethical fading to describe the process by which the moral dimensions of decisions become progressively less visible to the decision-makers who make them. Ethical fading occurs when decisions that have significant moral implications are framed, through the ordinary processes of organisational life, in terms that make their moral dimensions invisible. A decision about dust control at Wittenoom is framed as a cost management decision, not as a decision about how many workers will develop mesothelioma. A decision about barrier adequacy at Macondo is framed as a project management decision, not as a decision about whether 11 people will survive the night. The ethical content of the decision fades as the organisational framing, commercial, operational, and technical, comes to dominate the decision-maker's understanding of what they are being asked to decide.

Bazerman and Tenbrunsel (2011) have argued that ethical fading is not the exception in organisational life. It is the norm. Most decisions that produce organisational harm are not made by people who are consciously choosing to cause harm. They are made by people whose perception of the ethical dimensions of their decisions has been progressively eroded by organisational framing, by cultural normalisation, and by the gradual disappearance of the moral salience of the human beings who will be affected by those decisions from the decision-maker's immediate field of attention. This erosion is moral drift: the same process as Normalised Tolerance, operating on ethical standards rather than safety ones.

> *Moral drift does not feel like a moral failure from inside the organisation. It feels like professional management. It feels like making hard decisions in a complex commercial environment. It feels like the kind of realistic, unsentimental leadership that organisations need to survive and prosper. It is only from outside, in the aftermath of the harm it has produced, that its moral character becomes visible.*

The Architecture of Blame and Its Limits

The standard response to organisational harm, in law, in public discourse, and in the cultural frameworks through which societies make sense of disasters, is to identify the individuals who made the critical decisions and to assign blame to them. This is the architecture of blame: the institutional and cultural structure through which moral accountability for organisational harm is allocated to specific persons, expressed in the language of individual culpability, and resolved through the mechanisms of individual sanction.

The architecture of blame is not wrong. Individual accountability matters. The managers who made the decisions that degraded process safety management at Texas City, the executives who resisted adequate dust control at Wittenoom, and the leaders who failed to develop the mindful engagement with operational reality that their roles required at Chernobyl and Gretley and Longford – these people bear genuine moral responsibility for the decisions they made and the harm those decisions contributed to. The systemic account of how those decisions were produced does not exempt them from that responsibility.

But the architecture of blame is incomplete. It is incomplete in the same way that the proximate-cause focus of disaster inquiry processes is incomplete, because it identifies a subset of the morally responsible parties at the expense of the fuller set that a complete moral account would require. When the harm produced by an organisational disaster is traced to its systemic origins, to the five conditions and the processes that produced them, the moral accountability that the full systemic account identifies extends considerably further than the individuals who made the proximate decisions.

It extends to the senior executives who established the cost-reduction cultures that produced Production Gravity. It extends to the board members who approved the financial targets that those cultures were designed to meet. It extends to the regulators who set and enforced safety standards at levels that accommodated the commercial interests of the industries they regulated. It extends to the industry associations that shaped those regulatory standards. It extends, in some cases, to the political and economic systems that created the competitive environments within which the organisations operated. The architecture of blame, focused on the individual proximate decision-makers, captures a fraction of the moral accountability that the full systemic account reveals.

This observation is not an argument for the diffusion of accountability, for the conclusion that if everyone is accountable, no one is. It is an argument for the expansion of accountability: for the recognition that the moral responsibility for organisational harm is distributed across the system that produced the harm in proportion to each actor's capacity to have seen the conditions developing and their authority to have addressed them. Those at the top of the organisations examined in this book had the greatest capacity and the greatest authority. They bear the greatest accountability. But they are not the only ones.

The Wilful Blindness Question

The concept of moral drift leads directly to one of the most important and most difficult questions in the ethics of organisational disaster: the question of wilful blindness. At what point does the failure to see the conditions that are developing within an organisation cease to be a genuine failure of perception and become, instead, a choice not to see, a wilful maintenance of the ignorance that allows the conditions to continue without producing the discomfort of acknowledged moral responsibility?

This question is not hypothetical. It has been raised, in various forms, in the legal and regulatory proceedings that have followed several of the disasters examined in this book. The directors and executives of organisations that have caused serious harm have, in some jurisdictions and some cases, been found to have been wilfully blind to the conditions that produced that harm, to have had the means to know and to have chosen, through the ordinary mechanisms of institutional self-protection and personal risk management, not to know.

The Wittenoom case is the most direct example in the Australian context. The evidence before the courts in the asbestos liability proceedings established that CSR Limited (formerly the Colonial Sugar Refining Company) and ABA (Australian Blue Asbestos – a wholly owned subsidiary of CSR that operated the Wittenoom mine) had access, through their own dust measurement programmes and through the available medical literature, to information about the hazard of asbestos exposure at Wittenoom that was sufficient to support the conclusion that the conditions in the mine and mill were dangerous. The failure to act on that information, to implement the dust control measures that the evidence supported, cannot be fully explained by reference to the genuine epistemic limitations of the time. It required, at some level, a choice not to engage fully with the implications of what the evidence showed.

Sunstein and Vermeule (2001) and subsequent legal theorists have argued that wilful blindness, the deliberate avoidance of knowledge that would impose unwelcome obligations, is a form of culpable ignorance: ignorance for which the ignorant party bears moral and, in some jurisdictions, legal responsibility. In the organisational disaster context, wilful blindness operates through the mechanisms of moral disengagement and ethical fading described earlier in this chapter. The organisation does not formally decide to ignore the evidence. It creates, through the ordinary processes of organisational life, a culture and a set of information systems that make it structurally unlikely that the evidence will be fully assembled, accurately interpreted, and honestly communicated to the people who would have to act on it. The result is a form of institutional ignorance that has the practical effect of wilful blindness, even when no single individual has made a conscious choice to ignore a specific piece of information.

This concept of institutional wilful blindness is important for the legal and regulatory frameworks within which organisational safety accountability operates in Australia and internationally. The due diligence obligations imposed on officers of corporations under Australian work health and safety legislation, and the equivalent provisions in other jurisdictions, create a positive obligation to take reasonable steps to ensure that the organisation is meeting its safety obligations. Those obligations cannot be discharged by reference to the institutional ignorance that the five conditions produce. The officer who has not developed the mindful leadership practices required to penetrate that ignorance, who has accepted the reassuring picture that the formal systems construct without probing beneath it, may find that their acceptance of that picture is assessed against the standard of due diligence rather than against the standard of sincere personal belief.

The Ethics of the Five Conditions

Each of the five conditions of the Blindspot Model has an ethical dimension that the safety science analysis does not fully capture. Examining those ethical dimensions is important for understanding the full moral character of the decisions and organisational processes that produce the Blindspot.

Signal Attenuation and the Duty of Communication

Signal Attenuation, the filtering of safety-critical information before it reaches the decision-making level, represents, at its core, a failure of the duty of honest communication that responsible organisational membership requires. The individuals who filter safety concerns, who soften the report, who reframe the anomaly, who resolve the near-miss at the operational level without escalation, are making choices about what information reaches the people responsible for acting on it. Those choices are not ethically neutral, even when they are made in good faith within cultural contexts that make them locally rational.

The organisation that produces a culture of Signal Attenuation through reward systems that punish bad news, through professional norms that conflate complaint with inadequacy, and through management practices that treat the escalation of safety concerns as an indication of operational failure bears a specific ethical responsibility for the consequences of the filtration it has produced. It has created, through its own cultural choices, the conditions in which honest communication of safety-relevant information is structurally disadvantaged. The harm that results from the information that never arrives is, in an important sense, the harm of those cultural choices.

Normalised Tolerance and the Corruption of Standards

Normalised Tolerance represents the progressive corruption of the safety standards that organisations establish to protect the people in their care. The standards exist because the organisation has, at some point, assessed the risk that the standard is designed to manage and determined that a specific level of protection is required. When that standard is revised downward, through the normalisation process, without a corresponding reassessment of the underlying risk, the organisation is, in effect, reducing the protection it provides to the people at risk without their knowledge and without their consent.

The ethical character of this reduction is not altered by the fact that it occurs gradually and through processes of cultural accommodation rather than through a discrete decision. The workers at Wittenoom did not consent to the progressive reduction of the dust control standards that the normalisation of operational practice produced. The NASA crew of Challenger did not consent to the progressive revision of the O-ring safety standards that the normalisation of deviance embedded. The harm that resulted from the normalised conditions was harm done to people who were operating on the basis of a false understanding of the protection they were being provided.

Structural Secrecy and the Duty of Transparency

Structural Secrecy represents a failure of the organisational transparency that the management of major hazard risk requires. Organisations that operate high-hazard facilities owe a duty to the people who work in and around those facilities and to the communities that live in proximity to them to maintain the information systems that allow the complete picture of the facility's risk state to be assembled, assessed, and acted upon. When Structural Secrecy prevents that complete picture from ever being assembled, the organisation is operating with an opacity about its own risk state that, from an ethical standpoint, is not simply a management failure. It is a failure of the transparency obligation.

The Pike River case illustrates this with particular force. The workers who went underground at Pike River on 19 November 2010 were entitled to reasonable assurance that the organisation responsible for their safety had a complete and accurate understanding of the hazards they were being asked to work in. The Structural Secrecy that prevented that understanding from being assembled was not morally neutral. It was the product of organisational choices about management structure, about the integration of safety information, and about the investment in the risk management systems that would have been required to assemble the complete picture that those workers and their families had a legitimate interest in seeing made differently.

Production Gravity and the Commodification of Safety

Production Gravity, in its most ethically stark form, is the commodification of safety: the treatment of safety protection as a cost variable to be optimised against commercial objectives rather than as a non-negotiable requirement of responsible operation. When the question becomes not whether to maintain safe conditions but whether the cost of maintaining them is justified by the expected risk reduction, the organisation has placed a commercial value on the safety of the people in its care, a value that may, in specific budget decisions, be outweighed by other commercial considerations.

This commodification is the ethical core of the Wittenoom story, and it recurs, in less extreme forms, in every case in which Production Gravity has been identified as a significant factor in the development of disaster conditions. It does not always look like what it is. It looks like prudent financial management. It looks like the responsible stewardship of organisational resources. It looks, from inside the organisation, like exactly the kind of hard-headed commercial judgement that organisations need their leaders to exercise. From outside, from the perspective of the workers whose safety is being commodified, it looks like something different. And it is something different.

Leadership Detachment and the Abdication of Stewardship

Leadership Detachment is, in its ethical dimension, an abdication of stewardship: the failure of leaders to discharge the fundamental responsibility that their authority over others imposes on them. The leaders of high-hazard organisations are not simply managers of commercial enterprises. They are stewards of the safety of the people who work within those enterprises, and, in many cases, of the communities that live and work in proximity to them. That stewardship obligation does not discharge itself through the existence of formal safety management systems, or through the delegation of safety responsibility to specialist functions, or through the receipt of reassuring safety performance reports. It requires the active, engaged, probing awareness of the organisation's actual safety condition that mindful leadership demands.

The ethical weight of this stewardship obligation is not diminished by the structural conditions that make its discharge difficult. The hierarchy filter, metric dependency, cultural closure, distance and scale, and institutional capture described in Chapter 8 are real challenges to the exercise of mindful leadership. They do not eliminate the obligation. They define the challenge that the ethical exercise of leadership authority in high-hazard organisations requires leaders to meet.

The Moral Responsibility of the Organisation

The analysis of moral drift and ethical corrosion through the lens of the five conditions leads to a conclusion about organisational moral responsibility that goes beyond the accountability of individual leaders and decision-makers. Organisations, the institutional structures, cultures, and systems through which the decisions examined in this book were made, bear a form of moral responsibility for the harm they produce that is distinct from, and not reducible to, the aggregate of the individual moral responsibilities of their members.

French (1979) argued that corporations are moral persons, that they have, through their organisational decision-making structures, the capacity for intentional action in a morally relevant sense, and that they therefore bear moral responsibility for the outcomes of those actions in a way that cannot be fully explained by reference to the intentions and choices of their individual members. This view is contested in the philosophical literature, and the debate about corporate moral agency continues. But the disaster evidence examined in this book provides a specific and practically important argument in its support: the patterns of harm documented across these cases are not adequately explained by the moral failures of individual decision-makers. They require, for their full explanation, a concept of organisational moral agency, a recognition that organisations can develop, through their structures, cultures, and accumulated practices, a character that is itself morally significant, and that the harm produced by that character is, in an important sense, the harm of the organisation rather than merely the harm of its members.

This concept of organisational moral responsibility has practical implications that extend beyond the philosophical debate. It implies that the response to organisational harm must include not just individual accountability, not just the identification and sanctioning of the individual decision-makers whose choices contributed to the harm, but organisational accountability: the assessment and reform of the structures, cultures, and systems through which those decisions were made. It implies that the prevention of future harm requires not just the replacement of bad individual decision-makers with better ones but the transformation of the organisational conditions that make harmful decisions predictable regardless of who is making them.

It implies, in short, that the Blindspot Model is not just a safety framework. It is a moral framework, one that identifies the specific organisational conditions through which organisations develop the character that makes them capable of producing catastrophic harm without intending to. And it implies that addressing those conditions is not just a safety management obligation. It is an ethical one.

Beyond Blame: Towards Moral Accountability

The argument of this chapter can be summarised as follows. The systemic account of organisational disaster, the account this book provides through the Blindspot Model, does not reduce the moral accountability of the individuals who produced the five conditions and who made the decisions that the five conditions shaped. It expands and enriches that accountability by showing where it is actually located: not only in the proximate decision-makers identified by inquiry processes but also in the full set of people, at every level of the organisations examined, who made the choices that produced the conditions.

It also adds a dimension of accountability that the architecture of blame cannot capture: the accountability of organisations themselves, as moral entities, for the character they have developed and the harm that character has produced. That accountability requires a different kind of response from the individualised blame attribution that disaster inquiry processes typically generate. It requires organisational reform rather than individual sanction as its primary mechanism, the transformation of the conditions that produced the harm rather than the removal of the individuals who were present when those conditions reached their catastrophic conclusion.

Singer (2022) has argued powerfully that the framing of accidents and disasters as products of individual failure is itself a form of moral failure, that it allows the systems and organisations that produce harm to escape the accountability they deserve by focusing attention on the people who happened to be at the sharp end when the harm occurred. The Blindspot Model provides the analytical framework for the broader accountability that Singer's argument demands. It shows not just that individual decisions contributed to the disasters examined in this book but also how the organisations within which those individuals operated produced the conditions that made harmful decisions predictable and what those organisations and their leaders could and should have done differently.

That is the moral weight of the active foresight framework developed in the final part of this book. It is not a management tool. It is the practical expression of the ethical obligation that the evidence demands: the obligation of leaders and organisations that operate with significant potential for harm to develop the capacity to see that harm developing before it becomes irreversible and to act on what they see with the seriousness that the welfare of the people in their care requires.

The question this book has been asking is not why organisations cause harm. After ten chapters and the evidence of 50 years of disaster scholarship, the answer to that question is clear enough. The question is what those who lead organisations owe to the people whose safety depends on them, and whether they are meeting that obligation. The answer, in every case examined in this book, was that they were not. Not because they were bad people. Because they had stopped seeing. And stopping seeing, when people's lives depend on your capacity to look, is not a management failure. It is a moral one.

The Transition to Active Foresight

This chapter completes the analytical work of Parts II and III. The cases examined across these chapters have established, with the rigour that the evidence demands, how the five conditions develop, how they interact, how they produce the perfect Blindspot in their most dangerous combination, and how the moral accountability for the harm they produce is distributed across the organisations and individuals that create and sustain them.

The remaining question, the question that the final part of this book addresses, is what follows from this analysis for the leaders, safety professionals, and investigators who are reading it. The analytical framework is not its own purpose. It is in service of a practical objective: the development of the organisational capacity to see the danger that is developing, in time to interrupt it, before the incubation period reaches its catastrophic conclusion.

Part IV turns from analysis to application. Chapter 11 examines the evidence from organisations and leadership teams that have successfully interrupted the Blindspot pattern, demonstrating that the conditions identified in this book are not inevitable features of organisational life but controllable variables that can be managed differently. Chapter 12 provides the practical framework, the active foresight approach, through which leaders and organisations can assess, address, and continuously monitor all five conditions simultaneously.

The evidence examined in this book is, in many respects, a record of organisational failure. But it is also, if read with the right questions in mind, a detailed map of where the failures occurred and what would have been required to prevent them. That map is what Part IV uses. What it produces is not a guarantee of safety. Complex systems will always retain some capacity for catastrophic failure. What it produces is the organisational capacity to see clearly, the active foresight that is, ultimately, the only adequate response to the danger that the Blindspot creates.

Part IV

Active Foresight

Chapters 11 and 12

The evidence assembled in this book is not a record of inevitability. It is a map of where the failures occurred and what it would have taken to prevent them.

Part IV turns from analysis to application. The preceding three parts have established what the Blindspot is, how each of its five conditions develops and operates, and what the convergence of those conditions produces. Part IV addresses what can be done about it.

Chapter 11 examines the organisations and industries that have, to varying degrees and through varying mechanisms, developed genuine capacity to identify accumulating danger before it becomes catastrophic harm. Their experience is not a template, but it is evidence: evidence that organisational blindness to accumulating risk is not an inevitable feature of complex systems but a controllable variable.

Chapter 12 provides the Active Foresight Framework: a structured, evidence-based instrument through which leaders and safety professionals can assess, address, and continuously monitor all five conditions in their own organisations. It is the practical instrument toward which the rest of this book has been working.

DOI: 10.4324/9781042017539-15

Chapter 11

What Seeing Looks Like

Organisations That Interrupted the Pattern

Evidence from the organisations and industries that have developed genuine capacity to identify accumulating danger before it becomes catastrophic harm.

The Other Side of the Evidence

The preceding ten chapters have been, necessarily, a study in failure. The cases examined, Deepwater Horizon, Grenfell, Challenger, Columbia, Pike River, Longford, Texas City, Wittenoom, Chernobyl, Gretley, and the Herald of Free Enterprise, are cases in which organisations failed to see the danger they were accumulating and in which that failure produced catastrophic harm. The evidence from those cases is the foundation of the Blindspot Model, and it is evidence that demands to be taken seriously.

But it is not the only evidence available. Alongside the disaster literature, the record of what happens when organisations fail to see, there is a less prominent but equally important body of evidence: the record of what happens when organisations do see. When they identify the warning signals before they reach the critical threshold. When they interrupt the normalisation before it becomes irreversible. When they assemble the fragmented risk picture before the fragments combine into catastrophe. When they counteract the Production Gravity before it erodes the margins beyond recovery. When their leaders maintain the genuine operational engagement that keeps them connected to the reality of the risk they are managing.

This evidence is less dramatic than the disaster record. It consists, largely, of things that did not happen, accidents that were prevented, disasters that were averted, and incubation periods that were interrupted before they reached their catastrophic conclusions. It is harder to study, harder to document, and harder to narrate compellingly than the evidence of the failures. But it is no less real, and for the practical purpose of this book, the development

DOI: 10.4324/9781042017539-16

of organisational capacity to see developing danger, it is, in some respects, more valuable.

This chapter examines that evidence. It draws on the high-reliability organisations literature, on specific case examples from aviation, nuclear power, healthcare, and the process industries, and on the analytical frameworks developed by Weick and Sutcliffe, Hopkins, and others, to describe what organisations that have successfully interrupted the Blindspot pattern look like and what they do structurally, culturally, and in terms of leadership practice that distinguishes them from the organisations whose failures are documented in the preceding chapters.

The picture that emerges is not a picture of perfection. High-reliability organisations are not organisations that never make mistakes. They are organisations that have developed the specific capabilities, the seeing capabilities, that allow them to detect and correct errors and developing hazards before those errors and hazards reach the point of catastrophic consequence. Understanding those capabilities is the essential preparation for Chapter 12's practical framework.

High-Reliability Organisations: The Foundational Evidence

The concept of high-reliability organisations, organisations that operate complex, high-hazard systems with consistently lower rates of catastrophic failure than the technology and the operational environment would predict, was developed by a group of researchers at the University of California Berkeley in the late 1980s and early 1990s. Todd La Porte and Pamela Consolini's (1991) initial work, examining the operations of US Navy aircraft carriers, air traffic control systems, and nuclear power plants, established the empirical foundation: some organisations operating in genuinely high-hazard environments were managing to achieve safety performance levels that their technological complexity and operational tempo made, in conventional terms, highly improbable.

Weick and Sutcliffe's (2001) development of the high-reliability organisation concept into a practical framework identified five specific processes that characterise the cognitive and organisational practices of high-reliability organisations. These five processes, which Weick and Sutcliffe group under the general concept of collective mindfulness, are directly relevant to the Blindspot Model because they represent, in effect, the specific organisational capabilities that counteract each of the five conditions.

The first process is preoccupation with failure: the active, ongoing concern with the possibility that something is going wrong, even, especially, in the absence of visible evidence that it is. High-reliability organisations treat the absence of bad news not as evidence of safety but as a signal that their detection systems may not be capturing what is developing. This preoccupation with failure is the epistemic counterforce to Normalised Tolerance: it maintains, against the normalising tendencies of operational experience, the awareness that conditions which have not yet produced catastrophe may still be dangerous.

The second process is reluctance to simplify: the active resistance to the interpretive shortcuts that make complex situations manageable by reducing them to familiar categories. High-reliability organisations invest in the maintenance of nuanced, complex understandings of their operational environments, understandings that resist the simplification that Normalised Tolerance and the hierarchy filter both produce. This reluctance to simplify is the counterforce to both Signal Attenuation and Normalised Tolerance: it maintains the complexity of the risk picture against the organisational tendencies that reduce it to manageable but inaccurate simplicity.

The third process is sensitivity to operations: the active, continuous awareness of the gap between how the organisation intends to operate and how it actually operates. High-reliability organisations invest in direct, ongoing contact between their leadership and operational environments, not through formal reporting systems but through the kind of direct engagement that produces genuine knowledge of operational reality. This sensitivity to operations is the counterforce to Leadership Detachment and, in part, to Structural Secrecy: it maintains the connection between the senior decision-making level and operational reality that the hierarchy filter and structural distance systematically erode.

The fourth process is commitment to resilience: the active development of the organisation's capacity to detect, contain, and recover from errors before they cascade into catastrophic outcomes. High-reliability organisations do not assume that their prevention systems will stop all errors from occurring. They invest in the detection and recovery capabilities that allow errors that do occur to be caught and corrected before they compound. This commitment to resilience is the counterforce to the compounding effects of the perfect blindspot: it creates the redundant detection mechanisms that allow a failure in one condition to be caught by another.

The fifth process is deference to expertise: the organisational practice of allowing decisions to migrate to the level of the hierarchy where the relevant expertise resides, regardless of that level's formal seniority. High-reliability organisations have developed the cultural and structural mechanisms that allow the person with the most relevant knowledge to have the most relevant authority in the moment it is needed and that prevent the Production Gravity of formal hierarchy from overriding the expertise of the person who actually understands the risk. This deference to expertise is the counterforce to the hierarchy filter and to the Production Gravity that shapes how authority is exercised in non-high-reliability organisations.

High-reliability organisations do not achieve their safety performance by eliminating risk from their operational environments. They achieve it by developing the specific organisational capabilities, preoccupation with failure, reluctance to simplify, sensitivity to operations, commitment to resilience, and deference to expertise that allow them to see the risks that are present and to respond to them before those risks become irreversible.

Aviation: The Evidence from a Transformed Industry

Industry Context: *Commercial aviation has, over the past five decades, achieved the most dramatic improvement in safety performance of any major high-hazard industry. The fatality rate per passenger kilometre flown has declined by approximately 99% since the 1960s. This improvement has occurred while the volume of air travel has increased enormously and while the complexity of the systems involved, including aircraft technology, air traffic management, and airline operations, has grown substantially. Understanding how the aviation industry has achieved this improvement is directly relevant to the active foresight framework this book develops.*

The Crew Resource Management Revolution

The transformation of aviation safety began, in important part, with a recognition that the majority of fatal accidents in commercial aviation were not caused by technical failure or by the malfunction of aircraft systems, but by failures of human performance: failures of communication, decision-making, and situational awareness in the cockpit. The analysis of accident data in the 1970s established this pattern with sufficient clarity that the Federal Aviation Administration and the major airlines began, from the late 1970s onwards, a sustained programme of research, training development, and operational change aimed at addressing the human factors dimension of aviation safety.

The result was crew resource management training, a systematic approach to developing the communication, decision-making, and team coordination skills of flight crews that has become one of the most extensively studied and widely implemented safety interventions in any high-hazard industry. Helmreich et al. (1999) documented the evolution and effectiveness of CRM training across multiple decades and multiple cultures, demonstrating that well-designed CRM programmes produced measurable improvements in crew performance on the specific behaviours most closely associated with accident causation.

The relevance of CRM to the Blindspot Model is direct and important. CRM training addresses, at the cockpit level, several of the specific mechanisms through which the five conditions of the Blindspot operate. It addresses Signal Attenuation by developing the communication practices that ensure safety-relevant information reaches every crew member who needs it, including the captain, whose authority might otherwise produce the hierarchy-based filtering that makes captains the last to know what their crew is observing. It addresses Leadership Detachment by developing the CRM norm of authority with approachability, the cultural standard that a captain should be genuinely open to challenge and correction from other crew members, regardless of the authority gradient that formal hierarchy creates. It addresses Normalised Tolerance by maintaining, through regular training and recurrent assessment, the standards of cockpit performance against which deviations can be identified and corrected before they become embedded cultural norms.

Just Culture and the Safety Reporting Revolution

The second major transformation in aviation safety was the development of what Dekker (2007) has called just culture: the cultural and institutional framework that allows safety-relevant information to be reported honestly without triggering the punitive responses that Signal Attenuation produces in organisational cultures where bad news is professionally dangerous.

Aviation's safety reporting systems, the Aviation Safety Reporting System in the United States, the Confidential Human Factors Incident Reporting Programme in the United Kingdom, and equivalent systems in other jurisdictions are designed specifically to counteract the cultural filtering mechanism of Signal Attenuation. They provide pilots, controllers, engineers, and other aviation professionals with a mechanism to report safety concerns, near-misses, and operational errors without the professional consequences that would otherwise make such reporting professionally risky. The confidentiality protections these systems provide are not primarily about protecting individuals from accountability. They are about protecting the flow of safety-relevant information from the cultural filtering that formal accountability systems tend to produce.

The evidence regarding the effectiveness of these reporting systems is substantial. The Aviation Safety Reporting System receives more than 100,000 reports per year and has, since its establishment in 1976, generated a continuous flow of safety intelligence that has been used to identify emerging hazards, improve training programmes, and revise operational procedures before those hazards reach the threshold of serious incidents or accidents (NASA, 2023). This is active foresight in institutional form: a system specifically designed to ensure that the safety signals generated by operational experience reach the people responsible for managing the aviation system's risk state in a form that allows those signals to be assembled, analysed, and acted upon before they compound into catastrophe.

Threat and Error Management: Seeing What Is Developing

The most recent major development in aviation safety thinking, Threat and Error Management, developed by Helmreich and colleagues at the University of Texas Human Factors Research Project, provides a direct bridge between aviation's operational safety practice and the active foresight concept this book develops.

Threat and Error Management (TEM) is a framework for understanding and managing the safety of flight operations by identifying, in real time, the threats that are present in the operational environment and the errors that crew members are making in response to those threats, before those threats and errors compound into incidents or accidents (Helmreich et al., 1999). TEM is, in effect, a real-time active foresight framework: a structured approach to seeing what is developing in the operational environment and managing it before it reaches a point of irreversibility.

The TEM framework has been incorporated into line operations safety audits across multiple major airlines, providing systematic evidence about the threat and error patterns that characterise normal flight operations, not just the patterns that produce accidents and incidents but also the patterns present in everyday operations that have the potential to do so. This systematic monitoring of the developing risk landscape, in the absence of catastrophic outcomes, is one of the clearest examples available of how organised foresight can be institutionalised in a high-hazard operational environment.

Nuclear Power: The High-Reliability Organisation in Institutional Form

Industry Context: *The nuclear power industry presents one of the most extensively studied examples of the deliberate, sustained, institutional effort to develop and maintain the organisational capabilities of high-reliability organisations. The industry's safety improvement following the Three Mile Island accident in 1979 and, more ambiguously, the response to Chernobyl in 1986, involved the systematic development of industry-wide safety standards, peer-review programmes, and safety culture assessment tools that have produced measurable and sustained improvements in nuclear safety performance across much of the world.*

The Institute of Nuclear Power Operations

The Institute of Nuclear Power Operations (INPO), established by the US nuclear industry following the Three Mile Island accident, is one of the most significant examples of an industry-level response to the challenge of maintaining high-reliability performance across a large number of complex, high-hazard facilities operated by different organisations under different managements.

INPO's central mechanism is the systematic peer evaluation of nuclear plant operations, a programme through which teams of experienced nuclear professionals from across the industry conduct detailed assessments of the safety performance of individual plants, producing findings and recommendations that are shared across the industry. These peer evaluations are specifically designed to address the Structural Secrecy and Leadership Detachment that Hopkins (2014) has identified as key factors in the degradation of safety performance at nuclear facilities. They provide an external perspective on the organisation's safety condition, a perspective that can see what the internal perspective, shaped by Normalised Tolerance and Leadership Detachment, has learned not to see.

The effectiveness of INPO's peer evaluation programme in improving nuclear safety performance has been the subject of substantial research. Rees (1994) documented the development of INPO's regulatory role and its impact on nuclear safety culture across the US industry, finding that the combination of peer evaluation, industry-wide standards, and the institutional pressure that INPO's assessments created for underperforming plants produced measurable improvements in safety performance across the fleet. The mechanism through which this improvement occurred, the systematic exposure of plant leadership to an external perspective on the gap between their formal safety systems and their operational reality, is, in the language of this book, a systematic intervention against Leadership Detachment and Structural Secrecy.

Safety Culture Assessment: Institutionalising the Look Inside

The nuclear industry's development of safety culture assessment tools, methods for systematically evaluating the safety culture of nuclear facilities and identifying the cultural conditions that increase the risk of safety performance degradation, represents one of the most sustained and rigorous institutional efforts to develop the organisational equivalent of the diagnostic framework this book provides.

The International Atomic Energy Agency's safety culture framework, the US Nuclear Regulatory Commission's safety culture policy statement, and the industry-developed safety culture models of INPO and the World Association of Nuclear Operators all represent attempts to operationalise the concept of safety culture in ways that allow it to be systematically assessed, managed, and improved. The specific dimensions of safety culture that these frameworks identify, leadership engagement with safety, safety communication, problem identification and resolution, and the management of production pressure map directly onto the five conditions of the Blindspot Model.

The nuclear industry's experience with safety culture assessment provides important evidence about both the possibilities and the limitations of systematic safety culture monitoring as a tool for active foresight. The possibilities are significant: regular, systematic safety culture assessments, conducted by trained assessors using validated instruments, can detect the early warning signs of the conditions, Normalised Tolerance, Signal Attenuation, And Production Gravity, before they reach dangerous levels. The limitations are equally important: safety culture assessments are only as valuable as the organisational response they generate. An assessment that identifies concerning trends in safety culture is a warning signal. Whether it produces the corrective action it warrants depends on whether the Leadership Detachment condition has been sufficiently addressed to ensure that the warning is received and acted upon.

Healthcare: Learning to See Systemic Risk

Industry Context: *Healthcare has, over the past two decades, undergone a significant transformation in its approach to patient safety, one that bears directly on the challenge of organisational foresight. The publication of the Institute of Medicine's landmark report To Err Is Human in 1999, which estimated that between 44,000 and 98,000 Americans died annually from preventable medical errors, triggered a sustained programme of safety improvement that has drawn heavily on the concepts and tools developed in the aviation and process industries.*

From Blame to System: The Patient Safety Movement

The patient safety movement that emerged from the To Err Is Human report represents, in important respects, the most ambitious application of the systemic account of harm to a major industry outside the traditional safety science context. The report's central argument, that medical errors were overwhelmingly the product of system failures rather than individual incompetence, and that reducing them required system redesign rather than individual blame and sanction, was, at the time of its publication, genuinely radical within the medical profession's cultural framework.

The implementation of that argument across healthcare systems internationally has not been uniform or universally successful. Edmondson's (2018) research on psychological safety in healthcare settings demonstrates the persistence, in many clinical environments, of the cultural norms, hierarchy, professional status, and the implicit expectation of individual infallibility that produce Signal Attenuation in its most direct form: the unwillingness of junior team members to raise concerns about the decisions of senior clinicians, even when those concerns are directly relevant to patient safety. The transformation of those cultural norms has been, in healthcare as in every other industry, the hardest and most important dimension of safety culture change.

What the patient safety movement has achieved, in the settings where it has been most effectively implemented, is a demonstration that the cultural transformation required is possible. Edmondson's (2018) research on high-performing healthcare teams identifies psychological safety, the shared belief that the team environment is safe for interpersonal risk-taking, specifically including the raising of safety concerns, as the single most important differentiator between teams with high safety performance and those with low performance. Her findings directly replicate, in the healthcare context, the high-reliability organisation research of Weick and Sutcliffe: the organisations that see most clearly are the organisations that have created the cultural conditions in which honest communication of concern is genuinely safe.

Structured Communication and the Interruption of Attenuation

One of the most practically significant developments in healthcare safety has been the development and implementation of structured communication tools designed to counteract the Signal Attenuation that hierarchy and professional culture produce in clinical settings. Tools such as SBAR (Situation, Background, Assessment, Recommendation), developed originally in the US Navy submarine service and adapted for clinical use, provide a standardised framework for communicating safety-relevant information across hierarchical boundaries in ways that are assertive enough to penetrate the cultural filtering that informal communication is subject to.

The evidence on the effectiveness of structured communication tools in reducing clinical adverse events is not uniformly positive; implementation fidelity and organisational culture both significantly moderate their effectiveness. But the best evidence, reviewed by Leonard et al. (2004), demonstrates that in settings where structured communication tools are well implemented and culturally supported, they produce significant improvements in the quality and completeness of safety-relevant information transfer across the hierarchical boundaries where attenuation most commonly occurs. They are, in effect, an organisational technology for counteracting one specific mechanism of Signal Attenuation, the hierarchy filter, by providing a structured pathway through which operational-level safety concerns can reach decision-makers with the clarity and assertiveness required to produce action.

The Process Industries: Learning from Disasters to Build Foresight

Industry Context: *The major hazard process industries, oil and gas, chemicals, and mining, have, in the aftermath of the disasters examined in this book, developed increasingly sophisticated approaches to safety management that incorporate elements of active foresight. The development of major hazard risk management frameworks, bow-tie analysis, barrier-based thinking, and process safety management systems represents the industry's sustained attempt to develop the organisational capacity to identify and manage the conditions that precede catastrophic events.*

Barrier-Based Thinking and the Management of Incubation

The bow-tie model of major hazard risk management, which maps the threats that can produce a hazardous event, the barriers that prevent those threats from producing the event, and the barriers that mitigate the consequences if the event occurs, provides a structural framework for the kind of active foresight that this book advocates. When it is used not merely as a documentation tool but as a genuine operational management framework, when the barriers are real, are actively monitored, and are maintained to the standard that their safety function requires, the bow-tie model creates a structured mechanism for detecting the degradation of safety margins before that degradation reaches the point of catastrophic failure.

Hopkins (2014) has examined the use of bow-tie and barrier-based frameworks in the process industries, identifying the conditions under which they function as genuine foresight tools and the conditions under which they become documentation exercises that satisfy compliance requirements without actually managing the risk they are supposed to address. The distinction is, in essence, a distinction between organisations that have addressed the Leadership Detachment condition, whose leaders are genuinely engaged with the operational state of the barriers, and those that have not. The bow-tie is a tool. Its value as a foresight instrument depends on the leadership culture within which it is used.

The Contribution of Incident Investigation to Active Foresight

One of the most important and most underutilised sources of active foresight in major hazard organisations is the systematic analysis of the near-miss and minor incident record. The near-miss events that occur in the months and years before major disasters, the events that are present in the incubation period as warning signals, are, by definition, already in the organisation's incident records. The challenge is not collecting them. The challenge is analysing them with the analytical depth and the systemic orientation required to see, in the aggregate pattern of near-misses, the developing conditions that the individual events, taken alone, do not reveal.

Reason (1997) argued that near-misses are, in a very specific sense, more valuable than accidents as sources of safety learning, not because they are less serious but because they are more frequent, they provide more opportunities for analysis, and they reveal the pre-accident conditions at a stage when corrective action is still possible. An organisation that treats its near-miss record as a statistical input to its safety performance metrics has access to the warning signals that its near-miss record contains but is not reading them. An organisation that treats its near-miss record as a primary source of intelligence about the developing conditions in its operational environment is doing something qualitatively different. It is practising active foresight.

The practical challenge is the analytical depth required to extract the systemic intelligence from near-miss records. Near-miss investigation, as it is practised in most organisations, is orientated toward the identification and correction of the immediate cause of the near-miss rather than toward the identification of the systemic conditions that produced it. Changing that orientation, moving from proximate-cause investigation to systemic condition analysis, is one of the most important and most achievable changes that organisations committed to active foresight can make. The tools for that analysis are available; ICAM, 5 Whys, the Swiss Cheese model, and other structured investigation frameworks all provide pathways to systemic understanding when they are used with the analytical commitment they require. The organisational will to use them at the depth required is what distinguishes organisations that are developing genuine foresight from those that are performing its appearance.

The Common Characteristics of Organisations That See

Across the aviation, nuclear, healthcare, and process industry examples examined in this chapter, a consistent set of organisational characteristics can be identified that distinguishes organisations with genuine active foresight capacity from those without it. These characteristics are not a formula for organisational perfection. They are the evidence-based description of what seeing looks like and what organisations that have successfully interrupted the Blindspot pattern have in common.

Characteristic One: Leadership That Actively Seeks Discomfort

The organisations with the strongest active foresight capacity are, without exception, organisations whose senior leaders actively and persistently seek out the information that makes them uncomfortable: the bad news, the operational problems, and the gap between formal systems and operational reality. This is not a natural tendency of senior leadership in large organisations. It requires deliberate personal effort and deliberate institutional investment. Leaders who seek discomfort rather than reassurance are creating, through their behaviour, the cultural conditions in which honest communication of safety concerns is professionally valued rather than professionally risky. They are the most direct and most powerful counterforce to Signal Attenuation available within any organisation.

Characteristic Two: Dedicated Safety Intelligence Functions

The organisations with the strongest active foresight capacity consistently invest in dedicated safety intelligence functions: organisational capacities specifically charged with assembling the fragmented safety information that Structural Secrecy distributes across the organisation, analysing it for patterns that reveal developing conditions, and communicating those patterns to the leadership with the clarity and authority required to produce action. These functions are distinct from compliance functions; their purpose is not to verify that the organisation is meeting its regulatory obligations but to identify the conditions that the regulatory framework is not capturing.

The structural independence of these functions from the operational management hierarchy is a consistent feature of the organisations in which they are most effective. Hopkins (2019) argues that the placement of safety intelligence functions within the operational management hierarchy is, in itself, a form of Production Gravity: it subjects the safety function to the same commercial pressures that shape every other part of the operational hierarchy, compromising its capacity to identify and communicate the conditions that those pressures are producing. The organisations that have addressed this structural problem, that have created safety intelligence functions with direct reporting lines to the most senior level of the organisation and with the protected resources required to function independently, have the most reliable and the most useful active foresight capacity.

Characteristic Three: Learning Systems That Go Deep

The organisations with the strongest active foresight capacity invest in learning systems that go beyond the identification of proximate causes to the analysis of systemic conditions. Their incident and near-miss investigation processes are designed not just to prevent the specific failure from recurring but to identify the organisational conditions, the five conditions, that produced the failure and that, if unaddressed, will produce the next failure in a different but structurally similar form.

Dekker (2014) has argued that the dominant approach to incident investigation in most organisations, focused on root cause identification and corrective action directed at the identified root cause, is systematically inadequate for the purpose of identifying the developing conditions that precede major disasters. His advocacy for what he calls a just culture of learning, one that treats incidents and near-misses as sources of systemic intelligence rather than as occasions for individual accountability, is directly consistent with the active foresight approach this book develops. The organisations that have implemented this approach most effectively are the ones that treat their safety learning systems not as compliance exercises but as intelligence functions, and that invest the analytical depth required to extract from those systems the systemic understanding that the intelligence they contain can provide.

Characteristic Four: Cultures That Reward Challenge

Without exception, the organisations with the strongest active foresight capacity have developed the cultural conditions in which raising safety concerns, challenging the prevailing operational consensus, identifying deviations from safe practice, and questioning decisions that carry safety implications, are professionally rewarded rather than professionally penalised. This cultural condition is both the most important and the most difficult to establish and maintain.

Edmondson's (2018) research on psychological safety demonstrates that the cultural conditions that support honest safety communication are fragile; they are easily damaged by specific leader behaviours (the dismissal of a safety concern and the punishment of a bad news reporter), and they require sustained, active, visible reinforcement by senior leaders to be maintained. The organisations that have created and maintained these conditions are the organisations that have invested, consistently and visibly, in the signals that tell their people that raising concerns is valued, not just in policy statements and training programmes, but in the actual observable behaviour of their senior leaders when safety concerns are raised.

Characteristic Five: Chronic Unease as an Organisational Norm

The final characteristic shared by organisations with genuine active foresight capacity is what Hopkins (2007) calls chronic unease: the institutionalised, culturally embedded awareness that good safety performance is never a reason to relax vigilance, that the absence of recent accidents and incidents is not evidence of safety but evidence of the need to look harder for the conditions that accidents and incidents reveal. Chronic unease, maintained as an organisational norm rather than a personal quirk of individual leaders, is the cultural counterforce to Normalised Tolerance: it keeps alive, against the normalising tendencies of operational experience, the awareness that the conditions that will produce the next disaster may already be developing in ways that current monitoring is not capturing.

Maintaining chronic unease as an organisational norm requires specific and sustained effort, because it runs directly against the psychological tendencies that operational success produces. Success breeds confidence. Confidence, in organisational contexts, tends to erode the vigilance that chronic unease requires. The organisations that have maintained chronic unease as a genuine cultural norm are the organisations whose leaders have most consistently modelled the epistemic humility that acknowledges success as a reason for continued vigilance rather than as a reason for relaxation of it.

> *The organisations that see most clearly are not the ones with the most sophisticated safety management systems, or the largest safety departments, or the most comprehensive regulatory compliance programmes. They are the ones whose leaders have created the cultural conditions in which honest communication of danger is safe, whose safety intelligence functions have the independence and the resources to assemble the complete risk picture, and whose people have internalised the understanding that yesterday's safety performance is not a guarantee of tomorrow's.*

What the Evidence Demands

The evidence reviewed in this chapter establishes, beyond reasonable doubt, that the active foresight capacity required to interrupt the Blindspot pattern is achievable. It has been achieved, across multiple industries and multiple organisational contexts, by organisations that made the deliberate, sustained, resource-backed commitment to developing it. It is not a utopian aspiration. It is a documented organisational reality.

What the evidence also establishes is that achieving it is genuinely hard. It requires leadership behaviours that run against the grain of the cultural norms that large organisations tend to produce. It requires structural investments that compete with production-focused uses of the same resources. It requires cultural transformations that take years to achieve and that require continuous active maintenance to sustain. The organisations that have developed genuine active foresight capacity have not done so by adding a safety programme to an otherwise unchanged organisational culture. They have done so by making the kind of fundamental, sustained, leadership-driven commitments that the evidence shows are required.

Chapter 12 provides the practical framework for making those commitments, the structured, evidence-based approach to assessing and addressing all five conditions of the Blindspot Model that the evidence from this chapter, and from the preceding ten, supports. It does not promise that implementing the framework will guarantee safety. Nothing can guarantee that. What it promises is that organisations that implement it will have developed the genuine capacity to see the danger that is developing and to act on what they see with the seriousness and the timeliness that the welfare of the people in their care requires.

Chapter 12

A Practical Framework for Active Foresight

How to assess, address, and sustain the organisational capacity to see accumulating danger before it becomes catastrophic harm.

From Analysis to Action

The preceding eleven chapters have established, with the rigour that the evidence demands, what the Blindspot is, how it develops, how its five conditions interact, what it looks like when all five conditions converge, and what organisations that have successfully interrupted it have in common. That analytical foundation is necessary. It is not, on its own, sufficient. The purpose of this book is not to produce a better understanding of why disasters happen. It is to contribute to preventing them.

This chapter provides the practical framework for doing that. The Active Foresight Framework is a structured, evidence-based approach to assessing and addressing the five conditions of the Blindspot Model within any organisation operating in a high-hazard environment. It is not a checklist. It is not a compliance tool. It is a framework for developing the genuine organisational capacity to see accumulating danger before it reaches the point of irreversibility, the capacity that the cases examined in this book consistently lacked, and that the organisations examined in Chapter 11 have, in their different ways, managed to develop.

The framework is organised around five modules, one for each condition of the Blindspot Model. Each module contains three elements: a diagnostic assessment process for identifying whether and to what degree the condition is present in the organisation; an intervention framework specifying the structural, cultural, and leadership changes required to address the condition; and a monitoring approach for sustaining the improvement and detecting deterioration before it becomes dangerous.

A sixth element addresses the integration of all five modules into a coherent organisational programme, because, as Chapter 9 established, addressing

DOI: 10.4324/9781042017539-17

any single condition in isolation is insufficient when the conditions are operating in combination. The Active Foresight Framework is designed to be implemented as an integrated programme, though individual modules can be used independently for diagnostic or targeted improvement purposes.

Before setting out the framework, three preliminary points deserve emphasis. First, the framework is not a substitute for the analytical understanding developed in the preceding chapters. Leaders and safety professionals who want to apply it effectively need to understand what they are looking for, and that understanding requires the theoretical and evidential foundation that the book has provided. Second, the framework is not a one-time exercise. Active foresight is a continuous organisational capacity, not a periodic audit. The monitoring approaches in each module are as important as the diagnostic-and-intervention elements. Third, the framework requires genuine leadership commitment to be effective. It will not function as a management exercise conducted below the level of genuine senior leadership engagement. The evidence from Chapter 11 is unambiguous on this point (Figure 12.1).

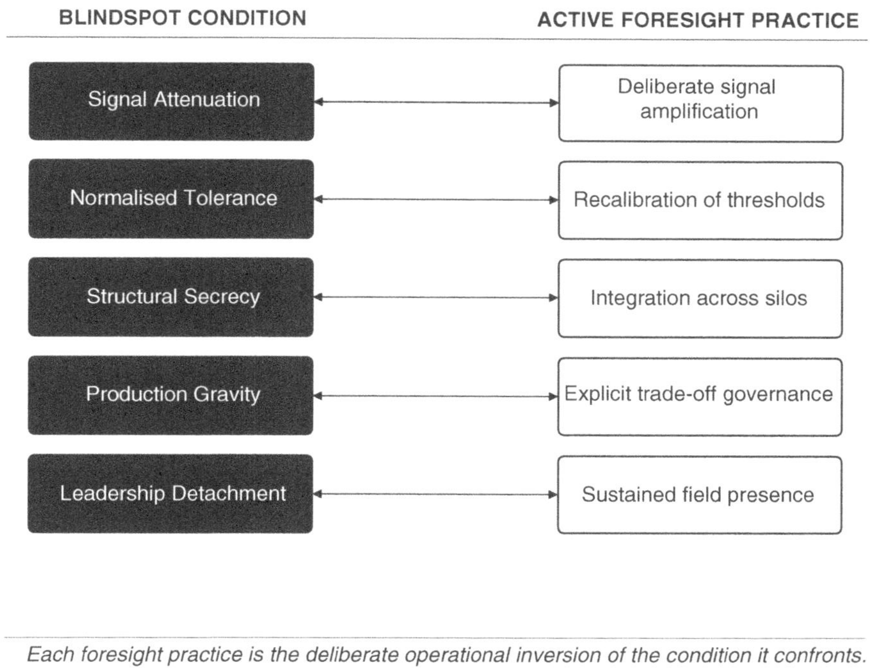

Figure 12.1 The Active Foresight Framework: pairing each Blindspot condition with its operational countermeasure.

Module One: Assessing and Addressing Signal Attenuation

Diagnostic Assessment

The diagnostic assessment for Signal Attenuation has three primary instruments, each targeting a different mechanism through which attenuation operates.

The first instrument is the communication gap survey: a structured, anonymised survey of operational personnel at multiple levels of the organisation, designed to establish the degree to which safety-relevant concerns are being raised, received, and acted upon at each level of the hierarchy. The survey asks operational personnel to report, anonymously, the frequency with which they have observed safety concerns at their level that were not escalated to the level above, their assessment of the reasons for that non-escalation, and their experience of the organisational response when concerns have been escalated. The gap between what the survey reveals about operational-level safety concerns and what the formal reporting systems have captured at the senior leadership level is the primary measure of Signal Attenuation in the organisation.

The second instrument is the incident trajectory analysis: a systematic review of the organisation's incident and near-miss records, examining the pathway of specific incidents from initial observation or report to final resolution or escalation. The analysis identifies the points in the reporting hierarchy at which incidents were resolved, the proportion that reached senior management review, and the characteristics of incidents that were resolved at the operational level without escalation. Incidents involving major hazard processes that were resolved without senior management review are a specific indicator of attenuation in the major hazard management domain, the domain where attenuation carries the highest risk of catastrophic consequence.

The third instrument is the leadership information audit: a structured assessment of the information available to senior leaders about the major hazard risk state of the organisation's operations. The audit asks senior leaders to describe, specifically and in operational detail, the current major hazard risk profile of the operations they are responsible for, what hazards are present, what barriers are in place, what the current state of those barriers is, and what the current near-miss and incident trends suggest about developing conditions. The gap between what leaders can describe and what the operational evidence shows is a direct measure of the Signal Attenuation that the hierarchy filter has produced.

Intervention Framework

Addressing Signal Attenuation requires intervention at three levels simultaneously: cultural, structural, and leadership.

The cultural intervention targets the norms that discourage the upward transmission of bad news. It requires visible, specific, and consistent leadership behaviour that demonstrates, in practice rather than in policy statements, that raising safety concerns is professionally valued. The most powerful single cultural intervention is the deliberate and public positive response by senior leaders to the first significant safety concern that is raised following the launch of the programme, a response that is visible, that specifically acknowledges the value of the concern raised, and that demonstrates the kind of substantive engagement with the concern's content that the cultural change requires.

The structural intervention targets the communication architecture that shapes how safety-relevant information travels through the organisation. It involves the review and redesign, where necessary, of the reporting structures, escalation protocols, and safety communication systems that define the pathways through which operational-level safety information reaches senior decision-makers. The key design principle is redundancy: the creation of multiple pathways through which safety-critical information can reach the senior level, so that the filtering of any single pathway does not prevent the information from arriving.

The leadership intervention targets the habits and practices of senior leaders directly. It involves the development, through coaching, peer exchange, and deliberate personal commitment, of the specific practices, regular operational site visits, direct worker engagement, and structured safety conversations that go below the level of formal reporting that create and maintain the genuine connection between the senior leadership level and operational reality that Signal Attenuation severs.

Key diagnostic question: If the most operationally experienced person in your organisation identified a serious safety concern today, what pathway would that concern travel to reach the most senior person with the authority to act on it, and what is the probability that it would arrive undiluted?

Monitoring Approach

The ongoing monitoring of Signal Attenuation requires three specific measures. The first is the repeat communication gap survey, conducted at regular intervals, annually at minimum, more frequently in organisations undergoing significant change, to track trends in the gap between operational-level concern and senior-level awareness. The second is the escalation rate monitor: a regular review of the proportion of significant safety events, near-misses, and hazard reports that are escalating to senior management review, tracking trends over time and investigating significant reductions in escalation rates as potential indicators of cultural attenuation. The third is the leadership information quality assessment: a periodic structured comparison of senior leaders' understanding of the operational risk state against the operational evidence, conducted by an independent function with access to both the senior leadership team and the operational environment.

Module Two: Assessing and Addressing Normalised Tolerance

Diagnostic Assessment

The diagnostic assessment for Normalised Tolerance requires a specific form of historical analysis that most organisations have not previously applied to their safety records: the longitudinal anomaly review.

The longitudinal anomaly review examines the organisation's record of known operational anomalies, recurring near-misses, equipment performance variations, procedural deviations, and process parameter exceedances over an extended historical period, typically five to ten years. For each anomaly, the review traces the history of the organisation's analytical response: how was it assessed when first identified? How has the assessment evolved over time? What standards have been applied to each assessment, and how do those standards compare to the original assessment and to the organisation's current formal safe operating parameters?

The signature of Normalised Tolerance in this review is the progressive relaxation of the standards applied to the same anomaly over time, driven by accumulated operational experience rather than by revised engineering analysis. An anomaly that was initially assessed as requiring urgent resolution but that has been progressively reclassified through a series of analyses, each building on the conclusions of the previous one, is an anomaly that may have been normalised. The longitudinal review makes this pattern visible in a way that cross-sectional safety reviews cannot.

The second diagnostic instrument is the operational standard alignment audit: a systematic comparison of the organisation's current operational practices, what is actually happening in the field, against the organisation's formal safe operating standards. The gap between the two is the measure of the normalisation that has occurred. This audit requires direct observation of operational practice, not review of documentation. Its purpose is to identify the specific practices that have drifted from the formal standard, the procedural shortcuts that have become routine, the operating parameters that are routinely exceeded, and the equipment conditions that are tolerated despite not meeting the maintenance standard, and to assess whether those deviations represent Normalised Tolerance or legitimately revised standards.

Key diagnostic question: For the three most significant operational anomalies you are currently managing, how does your current assessment of their acceptability compare to the assessment that was made when they were first identified, and what has driven the difference?

Intervention Framework

Addressing Normalised Tolerance requires the explicit re-engagement of the organisation's analytical processes with the historical record of anomaly assessment, using the original standards as the reference point rather than the most recent assessment. This is technically and culturally demanding. Technically, because it requires the assembly and systematic review of a historical record that organisations often do not maintain in a form amenable to longitudinal analysis. Culturally, because it requires the organisation to acknowledge that its current assessment of a known anomaly may be a product of normalisation rather than of genuine technical understanding, the experience it has relied on as evidence of acceptability may be misleading it.

The structural intervention for Normalised Tolerance involves the creation of a specific organisational function, a major hazard anomaly review board, or equivalent, responsible for maintaining the longitudinal analysis of significant operational anomalies and for ensuring that the standards applied to ongoing assessments are grounded in genuine engineering analysis rather than accumulated operational experience. This function needs to have access to the original engineering basis for the organisation's safe operating standards, the authority to require fresh analysis when the historical assessment trail shows signs of normalisation, and the structural independence from operational management to perform that role without being subject to the Production Gravity that shapes assessments within the operational hierarchy.

The leadership intervention requires senior leaders to explicitly and visibly reset the epistemic standard for anomaly assessment: to make clear, through their engagement with specific anomaly reviews and through the questions they ask of the technical analyses presented to them, that the standard for acceptability is the engineering evidence rather than the operational record. This is Hopkins' chronic unease in practice: the leadership insistence that good luck is not a safety standard and that the fact that an anomaly has not yet produced catastrophe is not evidence that it will not.

Monitoring Approach

The ongoing monitoring of Normalised Tolerance requires regular longitudinal anomaly reviews, structured reviews of the organisation's significant anomaly record conducted at regular intervals by the anomaly review function, with findings reported directly to the senior leadership. The monitoring also requires the establishment of specific triggers for fresh engineering analysis of longstanding anomalies: events that reset the analytical process regardless of the accumulated assessment history, including threshold exceedances, changes in operating conditions, new technical information, and the passage of defined time periods since the last independent analysis.

Module Three: Assessing and Addressing Structural Secrecy

Diagnostic Assessment

The diagnostic assessment for Structural Secrecy begins with a risk picture completeness assessment: an exercise in which the organisation's most senior safety professional, working with the senior leadership team, attempts to produce a complete and current description of the organisation's major hazard risk profile, every significant hazard, every barrier, the current status of every barrier, every significant near-miss and anomaly, and the trends in the organisation's risk condition over the preceding 12 months.

The exercise is not primarily about the content of the risk picture produced. It is about the process of attempting to produce it. The obstacles encountered in that process, the information that cannot be found, the systems that do not communicate with each other, and the questions that cannot be answered because the relevant knowledge is held by a team in a different part of the organisation are the map of the Structural Secrecy that exists within the organisation. Every gap in the completeness of the risk picture that the exercise reveals is a gap produced by a specific structural feature of the organisation's information management architecture.

The second diagnostic instrument is the knowledge dependency audit: a systematic identification of the safety-critical knowledge that is held by specific individuals or teams within the organisation, including knowledge that has been transferred from predecessor roles, knowledge that exists only in the institutional memory of long-tenured personnel, and knowledge that is distributed across multiple functional boundaries without any formal integration mechanism. The knowledge dependency audit reveals the Structural Secrecy that is produced not by system fragmentation but by the concentration of critical knowledge in specific people and roles, the Longford pattern, where the departure of resident engineers created a knowledge gap that the organisation had not recognised as dangerous until the knowledge was gone.

Key diagnostic question: Could you, right now, assemble a complete and accurate picture of your organisation's current major hazard risk state from information that is currently held in your organisation's systems, and if not, what is preventing you?

Intervention Framework

Addressing Structural Secrecy requires investment in information integration at the architectural level: the redesign of the organisation's safety information systems to create the cross-functional integration that structural secrecy's absence allows, and the mechanisms that allow fragmented safety information to be assembled into the complete risk picture that no single part of the organisation currently holds.

The specific structural interventions include the establishment of an integrated major hazard risk register that draws information from all relevant functional systems, operations, maintenance, engineering, safety, and contractor management and that is owned and actively maintained by a function with the authority and the resources to keep it current; the creation of cross-functional risk review processes that bring together the holders of the fragmented risk picture to assemble and assess it as a whole; and the development of knowledge management systems that capture and maintain the safety-critical knowledge that currently resides in individuals rather than in institutional systems and that ensure that knowledge survives the personnel changes that would otherwise destroy it.

The leadership intervention for Structural Secrecy involves the explicit ownership by senior leaders of the completeness of the risk picture. Leaders who treat their knowledge of the organisation's risk state as adequate, who accept the picture that the formal reporting systems construct without asking whether it is complete, are allowing Structural Secrecy to persist by not demanding its interruption. Leaders who ask, persistently and specifically, whether the picture they are being provided with is complete, who want to know what they do not know, and who invest the organisational resources required to find out, are the leaders whose engagement interrupts Structural Secrecy before it conceals a developing disaster.

Monitoring Approach

The ongoing monitoring of Structural Secrecy requires regular risk picture completeness assessments, conducted by the safety intelligence function, reported to senior leadership, and specifically designed to identify the dimensions of the risk picture that are currently incomplete and the structural reasons for that incompleteness. The monitoring also requires a knowledge retention programme that tracks the movement of safety-critical knowledge across the organisation, identifying, before personnel changes occur rather than after, the knowledge that would be lost and the steps required to retain it.

Module Four: Assessing and Addressing Production Gravity

Diagnostic Assessment

The diagnostic assessment for Production Gravity requires an approach that most safety management systems are not designed to support: the systematic analysis of the relationship between commercial and operational pressure and safety management decisions. This analysis is necessarily sensitive, and it requires a level of organisational candour about the role that commercial pressures play in operational decisions that many organisations consider difficult to achieve.

The first diagnostic instrument is the safety investment trend analysis: a longitudinal review of the organisation's investment in safety management, capital expenditure on safety-critical equipment and systems, operational expenditure on safety functions, staffing levels in safety-relevant roles, mapped against the organisation's commercial performance and against the safety performance outcomes over the same period. Sustained reductions in safety investment that correlate with periods of commercial pressure, without corresponding reductions in the major hazard risk profile of the operation, are a direct indicator of Production Gravity operating through the budget mechanism.

The second diagnostic instrument is the safety-production decision audit: a structured review of a sample of recent operational decisions in which safety and production considerations were in tension, decisions about maintenance scheduling, operating parameter management, contractor selection, capital allocation, and procedure compliance, examining the basis on which those decisions were made and the degree to which commercial considerations shaped their outcome. The audit is not designed to identify decisions that were wrong. It is designed to identify the decision-making culture and the institutional frameworks within which those decisions were made to reveal the degree to which the organisation's decision-making processes provide genuine protection against Production Gravity or whether they systematically favour production-positive outcomes in conditions of competing considerations.

Key diagnostic question: In the last 12 months, how many safety-motivated decisions have been made in your organisation that had a significant negative commercial consequence, and what does the answer to that question tell you about the actual balance between safety and production in your organisation's decision-making culture?

Intervention Framework

Addressing Production Gravity requires the most fundamental of the five condition interventions, because Production Gravity is rooted in the economic and operational imperatives that are, in a very real sense, the organisation's primary reason for existence. The intervention cannot eliminate production pressure. It must create, within the organisation's decision-making architecture, genuine structural countermeasures to Production Gravity that are robust enough to function under the commercial conditions in which the organisation actually operates.

The most important structural intervention is the separation of process safety governance from operational management. Hopkins (2019) argues that this separation is the single most significant structural determinant of whether major hazard safety management will be genuinely effective or whether it will be progressively compromised by the Production Gravity operating within the operational management hierarchy. The specific mechanism of this separation, whether through a dedicated process safety function with direct board reporting, through an independent safety case regime, through a major hazard regulator with genuine enforcement authority, or through some combination of these, will vary by industry and regulatory context. The principle is consistent: the people responsible for assessing and maintaining the adequacy of major hazard risk management cannot be subject to the same production pressures as the people responsible for operational performance.

The second structural intervention is the development of genuine process safety leading indicators, measures of the current state of the major hazard management systems that provide early warning of deterioration before that deterioration produces a serious incident. The distinction between personal safety lag indicators and process safety leading indicators, discussed in Chapter 7 in the context of the Texas City disaster, is central to this intervention. Organisations that are managing Production Gravity effectively have metrics that tell them, in real time, whether their major hazard barriers are intact and functioning, not metrics that tell them how many people were injured last quarter.

The leadership intervention for Production Gravity requires the explicit, visible, and consistent leadership message that safety is not a variable, that the commercial cost of maintaining safe conditions is a cost the organisation will always pay, and that the alternative is not commercially available. This message is not credible when it appears only in safety policy statements. It becomes credible when leaders make specific, visible decisions that incur a real commercial cost in order to maintain a safety standard and when those decisions are communicated to the organisation as the expression of a genuine value rather than a regrettable necessity.

Monitoring Approach

The ongoing monitoring of Production Gravity requires regular safety investment reviews, structured assessments of whether the organisation's investment in safety-critical systems and functions is maintaining the level required for the management of its current major hazard risk profile, conducted by the safety governance function and reported to the board. It also requires the ongoing tracking of process safety leading indicators, with defined response thresholds that trigger management review and corrective action when the indicators show deterioration below the level required for safe operation.

Module Five: Assessing and Addressing Leadership Detachment

Diagnostic Assessment

The diagnostic assessment for Leadership Detachment has three instruments, each targeting a different dimension of the condition.

The first instrument is the leadership safety knowledge assessment: a structured interview with each member of the senior leadership team, conducted by an independent assessor, in which leaders are asked to describe in specific operational detail the current major hazard risk profile of the operations they are responsible for. The assessment evaluates not just the accuracy of what leaders know, though accuracy matters, but the quality of their epistemic engagement with the question: whether they know how they know what they believe, whether they have tested the sources of their knowledge against operational reality, and whether they are aware of the dimensions of the risk picture that their current information sources do not capture.

The second instrument is the operational reality comparison: a structured comparison, conducted by independent safety professionals with direct operational access, between what senior leaders believe about the operational practice of their organisations and what is actually happening in the field. The comparison is not a gotcha exercise. It is a genuine attempt to map the gap between the leadership's understanding of operational reality and the operational reality itself, the gap that Leadership Detachment produces and that mindful leadership practices are required to close.

The third instrument is the leadership engagement audit: a systematic review of how senior leaders actually engage with the safety management of their operations, how often they visit operational sites, the nature of their interactions with frontline workers and supervisors, the questions they ask in safety reviews, and the degree to which they probe beneath the surface of formal reporting, compared against the engagement standard that mindful leadership requires. The audit is not about quantity of engagement alone. It is about the quality and the epistemic depth of the engagement: whether it is producing genuine knowledge of operational reality or merely creating the appearance of operational engagement without the substance.

Key diagnostic question: When did each member of your senior leadership team last have a direct, unscripted conversation with a frontline worker about the safety conditions of their work – not a town hall or a safety walk, but a genuine conversation in which the leader listened more than they spoke?

Intervention Framework

Addressing Leadership Detachment requires the development of specific leadership habits, structures, and cultural norms that maintain the genuine connection between the senior leadership level and operational reality. The intervention framework draws directly on the characteristics of mindful leadership identified by Hopkins and the high-reliability organisation practices identified by Weick and Sutcliffe.

The habit intervention involves the specific, scheduled, and genuinely unscripted direct engagement of senior leaders with operational environments and operational personnel. This is not the ceremonial safety walk in which a senior leader visits a site with a prepared schedule and a supporting cast of supervisors. It is the kind of direct, informal, operationally engaged conversation that produces genuine knowledge of what is actually happening in the field, the conversation in which a senior leader sits in the lunch room and asks the workers what worries them or walks through the maintenance bay and asks the technicians what they would fix if they could. These habits need to be scheduled, protected from being displaced by other demands, and treated by the leader's personal practice as essential rather than aspirational.

The structural intervention involves the creation of direct communication mechanisms between the operational frontline and the senior leadership that bypass the hierarchy filter: safety climate surveys reported directly to the board, anonymous safety concern reporting systems with direct senior leadership review, and structured operational feedback mechanisms that bring operational reality to the leadership table without passing through the layers of management that attenuate it. These mechanisms do not replace the hierarchy; they supplement it, creating the redundant information pathways that compensate for the filtering the hierarchy inevitably produces.

The cultural intervention involves the explicit modelling, by senior leaders, of the epistemic humility that mindful leadership requires: the visible acknowledgement that they do not know everything that is happening in their operations, that the formal picture they receive is incomplete, and that closing the gap between the formal picture and operational reality is a personal leadership priority rather than a function that can be fully delegated. Leaders who communicate this epistemic humility, not as weakness but as the honest acknowledgement of organisational complexity, create the cultural permission for the honest upward communication that both Signal Attenuation and Leadership Detachment require.

Monitoring Approach

The ongoing monitoring of Leadership Detachment requires regular leadership engagement audits, structured assessments of the quality and frequency of senior leadership engagement with operational environments, conducted by the safety governance function and reported to the board. It also requires regular operational reality comparison exercises, periodic independent assessments of the gap between leadership understanding and operational reality, conducted at intervals short enough to detect deterioration in leadership engagement before that deterioration produces the degree of detachment that the disaster cases demonstrate.

The Integration Module: Making Active Foresight an Organisational Capability

The five modules described above address each condition of the Blindspot Model individually. The integration module addresses the organisation's capacity to manage all five conditions simultaneously, continuously, and as an integrated system rather than as five separate programmes. This integration is what distinguishes genuine active foresight from the management of five separate safety improvement initiatives.

The Safety Intelligence Function

The organisational foundation of integrated active foresight is a dedicated safety intelligence function: a team specifically charged with assembling and maintaining the complete major hazard risk picture across all five conditions, analysing trends and patterns in that picture for evidence of developing Blindspot conditions, and providing the senior leadership and the board with the integrated risk intelligence they need to discharge their governance and due diligence obligations.

The safety intelligence function is distinct from the safety management function. Safety management is concerned with the implementation and maintenance of the organisation's safety management systems and the day-to-day operational task of ensuring that safety procedures are followed, that equipment is maintained, and that training is current. Safety intelligence is concerned with the state of the organisation's capacity to see developing danger, with monitoring the five conditions and providing early warning of their development before they compound into the perfect Blindspot.

The structural independence of the safety intelligence function is essential to its effectiveness. A function embedded in the operational management hierarchy will be subject to the Production Gravity that shapes every other part of that hierarchy; its findings will be moderated, its concerns will be softened, and its access to the senior leadership will be filtered by the same mechanisms that the function exists to counteract. The safety intelligence function needs direct reporting access to the most senior level of the organisation, to the CEO and, through appropriate governance structures, to the board, and it needs the protected resources and organisational authority to perform its role without being subject to the pressures that would compromise it.

The Active Foresight Review

The operational mechanism of integrated active foresight is the Active Foresight Review: a structured, regular senior leadership process specifically designed to assess the current state of all five Blindspot conditions, identify trends and patterns in the complete risk picture, and make the management decisions required to address conditions that are developing in concerning directions.

The Active Foresight Review is not a safety performance review. It does not primarily examine incident rates, near-miss counts, or compliance audit results, though these may be inputs to it. Its primary purpose is to examine the leading indicators of the five conditions: the communication gap measures, the anomaly assessment trends, the risk picture completeness, the safety investment levels, and the leadership engagement quality that indicate whether the organisation is maintaining the capacity to see or whether that capacity is deteriorating.

The review should be conducted at the senior leadership level, with the CEO and the senior leadership team, not with the safety management team in the CEO's absence, at intervals that are short enough to detect deteriorating trends before they become dangerous. For most high-hazard organisations, a quarterly review frequency represents a minimum. For organisations operating under significant production pressure or undergoing significant structural change, more frequent review is warranted.

The Board's Role

The board of any organisation operating in a high-hazard environment has a specific and important role in the integrated active foresight programme. That role is not the operational management of safety; that is the function of the executive team. It is the governance oversight of the organisation's capacity to see developing danger: the assurance that the five conditions are being systematically monitored, that the safety intelligence function has the independence and the resources to perform its role, and that the senior leadership is genuinely engaged with the active foresight programme rather than treating it as a management exercise that can be delegated below the level of genuine senior attention.

The board's discharge of this governance role requires specific information, information that is different from the safety performance information that most boards currently receive. A board that is receiving incident rates and compliance audit results is receiving information about what has already happened. A board that is receiving active foresight intelligence, information about the current state of the five conditions, trends in the developing risk picture, and the senior leadership's assessment of the actions required, is receiving information about what is developing. The difference between these two information sets is the difference between reactive governance and the kind of proactive governance that the management of major hazard risk requires.

Board-Level Governance Diagnostic: Ten Questions for the Board

The information set outlined above – intelligence about the organisation's current risk state, developing risk conditions, and the state of the organisation's capacity to see those conditions – is not conveyed well by standard safety dashboards. It requires a different conversation than the one that conventionally occurs between boards and executive teams. The following ten questions are offered as a structured means of initiating and sustaining that conversation. They are drawn directly from the five conditions of the Blindspot Model and are designed to distinguish between the receipt of safety performance information – what has already happened – and the receipt of active foresight intelligence – what is developing. They are a floor, not a ceiling: the beginning of a board's safety governance conversation, not its entirety.

1 On Signal Attenuation: What safety-critical information is reaching us through formal reporting channels, and what are we confident is being withheld, softened, or reframed before it arrives?
2 On Signal Attenuation: How do we know that frontline workers and first-line supervisors believe they can raise serious concerns upward without professional cost?
3 On Normalised Tolerance: Which current operating conditions or exemptions from our safety standards have we accepted because they have not yet produced harm, and how long has that acceptance been in place?
4 On Normalised Tolerance: When we reviewed near-miss and precursor event data in the last 12 months, how many signals did we reclassify as anomalies rather than as evidence of accumulating risk, and who made that judgement?
5 On Structural Secrecy: Where in our organisation are the people who each hold part of the picture of a developing hazard, and what mechanism ensures those parts are being assembled into a whole?
6 On Structural Secrecy: When did our safety, engineering, operations, and commercial leadership last examine the same risk information together, with the specific task of identifying what each could see that the others could not?
7 On Production Gravity: When safety considerations and production considerations have come into explicit conflict in our operations in the last quarter, how have those conflicts been resolved, and what pattern do those resolutions show?
8 On Production Gravity: Are the performance incentives of our operational leaders configured in a way that makes delaying production for a safety-relevant reason professionally rational or professionally costly?
9 On Leadership Detachment: When a member of this board last spoke directly with a frontline worker about safety, what did they learn, and how has that learning affected a decision we have made?
10 On Leadership Detachment: If our senior executives are consistently reassured about safety performance when they visit operational sites, what is the evidence that those reassurances are accurate rather than managed?

These questions are designed to produce discomfort. If they are asked regularly, and if the board insists on substantive answers, the condition of the organisation's active foresight capability will become visible at board level, which is where the ultimate accountability for that capability resides. If they produce consistently confident, comfortable, and reassuring answers, the board should treat that consistency as itself diagnostic – because it is consistent confidence in the absence of chronic unease that the scholarship surveyed in this book has repeatedly identified as a precursor to catastrophic harm.

> *Active foresight is not a safety programme. It is an organisational capability, the capability to see, accurately and completely, the danger that is developing within the organisation in time to act on what is seen before it becomes irreversible. Developing that capability requires investment, commitment, and sustained senior leadership engagement. It also requires the intellectual honesty to acknowledge that the organisations examined in this book, the organisations that did not see, were not, in most respects, unlike the organisations that employ the people reading it. The conditions they produced were ordinary. The capacity to see them is not natural. It requires work. This chapter has described that work.*

What the Absence of Disasters Does Not Tell You

The Blindspot Model and the Active Foresight Framework it supports are, by their diagnostic nature, instruments for identifying the conditions that precede failure. They are Safety-I instruments, in the terms developed by Hollnagel, Wears and Braithwaite (2015): they are concerned with understanding and interrupting the organisational processes by which things go wrong. That focus is deliberate and, this book has argued, necessary. It is also partial.

Hollnagel and his colleagues drew a distinction, in articulating the Safety-II perspective, between two very different ways of thinking about safety. The Safety-I perspective, which has dominated the safety management literature for most of its history, treats safety as the absence of accidents and studies accidents in order to understand how to eliminate their causes. The Safety-II perspective, by contrast, treats safety as the presence of successful performance under varying conditions and studies what makes that success possible. The two perspectives are not in opposition. They are complementary. A mature safety practice requires both.

The distinction matters for the practitioners to whom this book is addressed, because the tools described in the preceding modules are all directed at the Safety-I question: the question of what is going wrong that will, if uninterrupted, produce catastrophic harm. That question is essential. It is not the whole question.

The Safety-II question is, how does this organisation, most of the time, in most of its operations, succeed in keeping people safe under conditions that are variable, imperfect, and often imperfectly specified? What are the adjustments that operators, supervisors, teams, and managers are making, on a daily basis, that allow work to proceed safely despite gaps in procedures, fluctuations in resources, and unexpected disturbances? What, in other words, are the organisation's ordinary successes – and what makes those successes robust or fragile?

The practical importance of the Safety-II question follows directly from the earlier analysis in this chapter about the limits of lag indicators. An organisation that measures its safety performance exclusively through the absence of disasters learns nothing about the presence of safety: nothing about whether the routine successes of its operations are the product of a resilient system or of a system that is succeeding by accident, in ways that will not continue when conditions change. The absence of disasters is a necessary but not sufficient indicator of safety. An organisation can be accident-free and still be dangerous, if the absence of accidents is produced by luck, by the absorption of risk by the people at the sharp end, or by the combination of conditions that has not yet, but will eventually, produce harm.

For leaders using this framework, the practical implication is that the diagnostic questions of the preceding modules should be paired with a second set of questions directed at understanding how the organisation's ordinary work succeeds. Which adjustments do our operators make that compensate for gaps in our formal procedures? Where in our operations is the distance between work-as-imagined and work-as-done, and what does that distance tell us about where our systems are quietly fragile? When we experience a near-miss that does not produce harm, what was it that intervened to prevent the harm, and can we count on that intervention to operate again? These questions – the Safety-II questions – are what the Blindspot Model does not, by itself, answer.

The point is not that the Active Foresight Framework is inadequate. It is that it is deliberately bounded. It does what Safety-I instruments do, and it does it in an integrated way that the preceding literature has not. What the leaders who use it must hold in mind, simultaneously, is that their work is not finished when they have identified and interrupted the conditions for failure. Real foresight requires both: the capacity to see the conditions for harm, and the capacity to see the conditions for success, with the same seriousness and the same systematic attention.

From Seeing to Acting: Why Perception Does Not Always Produce Action

The diagnostic framework offered in this chapter is designed to make the conditions of organisational blindness visible. But visibility is not the endpoint. A condition that has been diagnosed but not acted upon is, in practical terms, indistinguishable from a condition that has not been diagnosed at all. The question this section addresses is why organisations that have, in fact, seen accumulating danger so frequently fail to treat what they have seen as action-demanding (Figure 12.2).

The scholarship on judgement and decision-making under uncertainty offers considerable help here. Kahneman's (2011) synthesis of several decades of research on cognitive bias and dual-process reasoning, together with Tversky and Kahneman's (1974) earlier work on the heuristics that underlie decisions in conditions of uncertainty, establishes that human decision-making does not operate by the orderly integration of evidence that classical decision theory assumes. It operates through fast, intuitive, pattern-recognition processes that are typically adaptive but that are, in specific circumstances, systematically biased in ways that produce predictable failures of judgement. Klein's (1998) complementary work on naturalistic decision-making, developed through field studies of experts making rapid decisions in high-consequence environments, shows that the pattern-recognition processes experts use are remarkably effective in familiar conditions but fail in characteristic ways when the situation under examination has features that the expert's pattern repertoire does not accommodate.

The intersection of these two literatures with the analysis of this book identifies three specific barriers that operate, in practice, to prevent organisational perception of developing danger from producing organisational action on that danger.

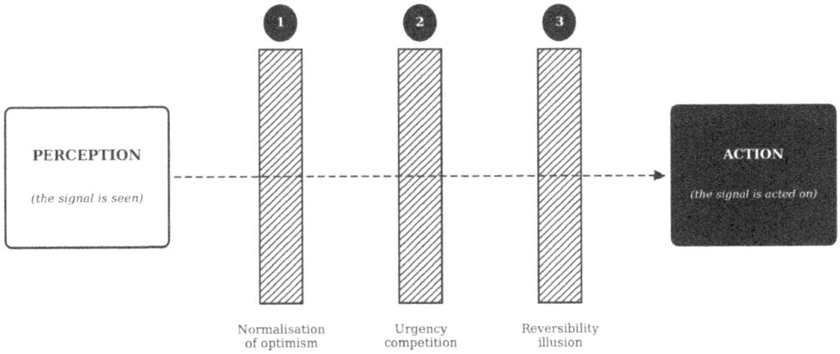

Figure 12.2 From seeing to acting: the three barriers that separate perception from organisational response.

The first barrier is the normalisation of optimism: the systematic organisational tendency to interpret bad news as an outlier, a temporary variation, or a problem that is on the verge of being resolved, rather than as evidence of an accumulating condition that requires action. The normalisation of optimism is related to, but distinct from, the normalisation of deviance examined in Chapter 5. Where normalisation of deviance operates on the perception of signals before they are consciously registered, normalisation of optimism operates on the interpretation of signals after they have been registered. It is the cognitive and social process by which a leader who has been shown a concerning pattern says, reasonably and in good faith, that it is probably an anomaly, that the situation is likely to resolve, and that no immediate action is required. Kahneman's work on the availability heuristic and on optimism bias explains why this response is so common: the reference class of 'concerning patterns that resolved without consequence' is large and readily recalled, while the reference class of 'concerning patterns that preceded catastrophic harm' is, for any given leader, smaller, more distant, and harder to retrieve at the moment of decision. The practical countermeasure is the deliberate construction of an institutional reference class: the regular, structured review of historical case evidence, drawn from this book's scholarship and from the organisation's own record, that keeps the reference class of consequential concerning patterns cognitively available.

The second barrier is the urgency competition problem: the systematic displacement of long-horizon safety concerns by short-horizon operational demands. A leader asked on any given day to choose between addressing a concern about an accumulating condition that has not yet produced harm and addressing a concern about an operational issue that is currently producing measurable cost or delay will, in the absence of specific countermeasures, consistently choose the latter. This is not a failure of judgement in the individual case. The operational issue has a defined timeline, a measurable consequence, and an organisational audience demanding attention; the accumulating condition has an undefined timeline, a hypothetical consequence, and no audience outside the safety function demanding that it be addressed today. Klein's naturalistic decision-making research identifies this as a characteristic pattern: under time pressure and competing demands, decision-makers allocate attention to the problems that present as immediate and defer attention to the problems that present as eventual. The practical countermeasure requires the structural elevation of active foresight intelligence to a status that competes with current operational demand: standing agenda items, protected time, and review processes that cannot be displaced by short-horizon pressure without a decision at the level of the board being recorded as having made that choice.

The third barrier is the reversibility illusion: the persistent cognitive bias toward believing that there will be a later opportunity to act on information that, in fact, requires action now. The reversibility illusion is particularly damaging in the context of accumulating organisational risk because the structural feature that makes the risk consequential – the fact that the conditions are compounding – is precisely the feature that the illusion suppresses. A leader who defers action on a developing condition implicitly models the condition as static, waiting for attention. The condition is, in fact, continuing to develop, and the window within which intervention would have been effective is continuing to close. The catastrophic outcome, when it arrives, arrives because the action that would have prevented it, while still available at the moment of deferral, became unavailable at some point thereafter, without the leader having been informed that the window had closed. The practical countermeasure is the explicit identification, in active foresight reviews, of the temporal character of each condition under consideration: whether it is stable, improving, or deteriorating; whether the window for effective intervention is widening or narrowing; and whether the deferral of action is, in effect, a decision to act later from a materially worse position than the one available now.

These three barriers operate together, and they operate in the same individuals and the same organisations that have, through the diagnostic apparatus of the preceding modules, successfully seen the conditions of developing harm. The work of building active foresight is not finished, in other words, at the point at which the conditions are visible. It is only finished at the point at which the visible conditions are translated into action before the window for effective action has closed. Seeing is not enough. Action, on what has been seen, at the time it must be taken, is what distinguishes the organisations that interrupt the pattern from the organisations that will, in due course, add their own case to the literature.

A Note on Implementation

The Active Foresight Framework is ambitious. Fully implementing all five modules and the integration programme across a large, complex organisation is a significant undertaking, one that requires genuine senior leadership commitment, sustained resource investment, and the willingness to confront the dimensions of the organisation's current safety condition that the diagnostic assessments will inevitably reveal.

For organisations that are beginning this work, a phased implementation approach is both practical and evidence-consistent. The evidence from Chapter 11 suggests that the most productive starting point is typically Module Five, Leadership Detachment, because addressing Leadership Detachment creates the organisational conditions under which the other four modules can be implemented most effectively. A senior leadership team that has genuinely engaged with the Leadership Detachment diagnostic, and that has made the personal and institutional commitments required to address it, is a leadership team that will drive the implementation of the other four modules with the genuine engagement required to make them effective.

The second priority is typically Module One, Signal Attenuation, because the direct communication improvements that addressing Signal Attenuation produces generate, in relatively short timeframes, the kind of operational intelligence that demonstrates the value of the active foresight approach to the broader organisation. When the communication gap survey reveals the concerns that the formal reporting systems have not been capturing, and when those concerns are addressed with the visible senior engagement that the cultural intervention requires, the message that active foresight is producing genuine value, rather than generating compliance paperwork, becomes credible to the people whose engagement is required to sustain it.

The remaining modules can be implemented in the sequence that best fits the organisation's specific risk profile, resource availability, and the diagnostic findings of the initial assessments. What matters is not the sequence. What matters is that all five modules are addressed, that their integration is pursued as a programme objective rather than as a by-product of separate initiatives, and that the senior leadership engagement required to make the programme effective is maintained with the consistency and the depth that the evidence from both the disaster cases and the high-reliability organisations demonstrates is required.

The organisations that see most clearly are not the ones that have been lucky. They are the ones that have done this work, that have invested, in advance of catastrophe, in the organisational capacity to look honestly at what is developing within them and to act on what they see. That investment is what this book is ultimately about. And making it, genuinely, completely, with the seriousness that the welfare of the people in your care demands, is what active foresight means in practice.

Epilogue

The Responsibility of Foresight

What the evidence requires of those who lead organisations with significant potential for harm.

This book began with a question about seeing. About the capacity of organisations, and the people who lead them, to see the danger that is accumulating within them before that danger reaches the point of irreversibility. It has examined that question through the evidence of 11 major industrial disasters, through four decades of disaster scholarship, through the experience of industries that have developed genuine capacity for organisational foresight, and through a framework that attempts to synthesise all of that into something practically usable.

What the evidence shows is not comfortable. It shows that the conditions producing catastrophic harm are not aberrations, not the products of exceptional dysfunction or exceptional malice. They are the outputs of ordinary organisational processes, operating under ordinary pressures, producing ordinary patterns of blindness that are, in the general case, entirely preventable. The disasters examined in this book were not inevitable. They were the products of conditions that could have been identified and interrupted. They happened because they went unseen.

That finding imposes an obligation. If the conditions that produce catastrophic harm are predictable and preventable, if the scholarship assembled in this book is right about that, and the evidence strongly suggests it is, then the failure to develop the organisational capacity to see those conditions developing is not a misfortune. It is a choice. Not usually a conscious choice, not usually a deliberate decision to accept known risk at the expense of known people. But a choice nonetheless: the choice to manage organisations in ways that produce and sustain the five conditions, rather than in ways that identify and interrupt them.

That choice is made every day, in organisations of every kind, by leaders who would describe themselves, accurately, in most cases, as genuinely

DOI: 10.4324/9781042017539-18

committed to the safety of the people in their care. It is made through budget decisions that defer safety investment. Through reporting cultures that filter the bad news before it reaches the top. Through the comfortable acceptance of the reassuring picture that formal systems construct, without the uncomfortable probing that would reveal what the picture is missing. Through the gradual, invisible accommodation of known hazards into the revised understanding of what is normal. Through the structural designs that fragment the risk picture across so many boundaries that no one person ever sees it whole.

The leaders who made these choices in the organisations examined in this book were not, in most cases, bad people. They were people operating within organisational systems that had been designed, through the accumulated decisions of many people over many years, to produce those choices rather than different ones. That contextual understanding is important for the accuracy of the causal account this book provides. It does not reduce the moral weight of what their choices produced.

> *The question that the evidence of this book poses is not whether you are committed to safety. Most leaders reading it are. The question is whether your commitment is sufficient, whether it is specific enough, structural enough, and honest enough about the conditions that ordinary organisational life produces to actually see the danger that is developing before it becomes irreversible. Commitment without foresight is not enough. The disasters documented in this book were full of committed people.*

What Foresight Requires

The active foresight framework developed in Chapter 12 is a practical tool. But its implementation requires something more than a structured programme. It requires a specific quality of leadership that this book has been circling throughout: the quality of genuine epistemic honesty about the organisation's actual condition.

Epistemic honesty, in this context, means the willingness to know what you do not know. To acknowledge that the picture your formal systems construct for you is incomplete. To treat the absence of bad news not as evidence of safety but as a reason to look harder. To resist the institutional tendencies, the cultural closure, the metric dependency, and the comfortable reassurance of hierarchical reporting that make it easy to believe the organisation is safer than it is. To ask the questions that the organisation's culture finds uncomfortable, and to create the conditions in which honest answers to those questions can be given.

This quality is not natural. It runs against the grain of the psychological tendencies that success produces, the confidence, the institutional familiarity, the accumulated experience that feels like knowledge but may be the very thing that has made the danger invisible. Developing and maintaining it requires deliberate effort: the specific habits, structures, and cultural commitments that Chapter 12 describes, sustained against the organisational tendencies that constantly push in the opposite direction.

It also requires courage. The courage to hear what operational personnel actually think about the safety of their work. The courage to commission and receive the findings of honest safety intelligence assessments. The courage to make the resource decisions that Production Gravity resists. The courage to acknowledge, publicly and specifically, when the organisation's formal safety picture and its operational reality do not match, and to treat that gap not as a threat to the organisation's reputation but as the most important safety information the organisation has.

That courage is not heroic. It does not require extraordinary personal qualities. It requires the ordinary determination to discharge the ordinary obligations of leadership in organisations that carry significant potential for harm. Those obligations have always been present. The evidence assembled in this book makes them harder to avoid seeing.

The Responsibility of the Profession

The argument of this book is addressed primarily to the leaders and safety professionals who are directly responsible for the safety of organisations operating in high-hazard environments. But it has implications that extend beyond the individual organisation to the broader professional community, the safety professionals, the regulators, the researchers, the educators, and the legal professionals, whose work shapes the institutional environment within which individual organisations manage major hazard risk.

The safety profession has, over the past two decades, made significant progress in developing the theoretical and practical frameworks for understanding and managing major hazard risk. The concepts this book draws on, Turner's incubation, Vaughan's normalisation of deviance, Dekker's drift into failure, Hopkins' mindful leadership, and Weick's collective mindfulness, represent a body of intellectual work of genuine quality and genuine practical value. The challenge is not the quality of the knowledge. The challenge is its translation into the ordinary practice of safety management in the organisations that need it most.

That translation challenge is a professional responsibility. Safety educators who teach compliance frameworks without teaching the systemic account of how organisations produce the conditions for catastrophic harm are producing safety professionals who know how to manage safety systems but not how to see when those systems are not doing what they are supposed to do. Safety regulators who assess compliance without assessing the five conditions are providing assurance that may, as the Texas City case demonstrates with devastating clarity, be entirely disconnected from the actual major hazard risk state of the organisations they regulate. Safety researchers who produce scholarship that never reaches the people responsible for implementing its insights in operational environments are generating knowledge that sits unused while the conditions it describes continue to develop.

The responsibility of foresight extends, in other words, to the institutions that shape what safety practice looks like in the organisations where disasters develop. The framework developed in this book is one contribution to closing that gap. It will not close it alone. What is required is the sustained, collective commitment of the safety profession, in its research, its education, its regulatory practice, and its organisational consulting, to the translation of the evidence into the kind of genuine active foresight capability that the evidence demands.

The Last Word

The men and women who died in the disasters examined in this book went to work. They did what people in high-hazard industries do every day; they showed up, they did their jobs, and they trusted that the organisations responsible for their safety had managed, with adequate competence and adequate care, the risks that those jobs involved. That trust was, in every case, misplaced. Not because the organisations were staffed by people who did not care, but because those organisations had developed the specific conditions, the five conditions, that made the honest, complete, accurate management of the risks their people were carrying structurally impossible.

This book has been written in the belief that those deaths were preventable. That the organisations that produced the conditions for them could have seen those conditions developing and could have interrupted them before they became irreversible. That the tools for seeing, the analytical frameworks, the organisational practices, and the leadership habits were available, or are now available, and that their application is a genuine and achievable obligation of every person who leads or advises or regulates an organisation with significant potential to harm the people within it.

The responsibility of foresight is not complicated. It is the responsibility to look, honestly, systematically, and with the specific analytical depth that the evidence requires, at what is developing within the organisations we are responsible for. To create the conditions in which that looking is possible. To act on what the looking reveals. And to maintain, against every organisational tendency toward the comfortable reassurance of the picture that formal systems construct, the chronic unease that genuine safety requires.

The disasters documented in this book did not have to happen. The next one does not have to happen either. What is required is the willingness to see.

Glossary

Key Terms and Concepts

The following glossary defines the key terms and concepts used in this book. Where a term originates from a specific scholar or framework, the source is noted. Terms developed specifically for this book are indicated as such.

Active Foresight: The organisational capacity to identify the conditions that precede catastrophic harm, the five conditions of the Blindspot Model, in sufficient time to interrupt them before they produce irreversible consequences. Active foresight is distinguished from reactive safety management by its prospective orientation: it is concerned with what is developing, not with what has already happened. Developed in this book.

Acceptable Risk Process: The formal procedure used by NASA and other high-hazard organisations to assess engineering anomalies and determine whether they are within acceptable risk parameters for continued operations. Vaughan (1996) identified the Acceptable Risk Process as a key mechanism through which normalisation of deviance operated within NASA during the shuttle programme.

Architecture of Blame: The institutional and cultural structure through which moral accountability for organisational harm is typically allocated to specific individuals, expressed in the language of individual culpability, and resolved through mechanisms of individual sanction. The architecture of blame is analytically insufficient because it focuses on proximate decision-makers rather than the full set of individuals and organisational conditions responsible for producing the harm. Developed in this book.

Blindspot Model: The original analytical framework developed in this book, identifying five interlocking organisational conditions, Signal Attenuation, Normalised Tolerance, Structural Secrecy, Production Gravity, and Leadership Detachment, that, when operating in combination, reliably produce the inability to see accumulating danger in time to prevent catastrophic harm. Developed in this book.

Bow-Tie Model: A risk management framework that maps the threats that can produce a hazardous event (the left-hand side of the bow-tie), the preventive barriers that stop threats from producing the event (the knot), and the mitigative barriers that reduce the consequences once an event has occurred (the right-hand side). Widely used in major hazard industries for risk assessment and safety management.

Chronic Unease: The sustained, active concern that the organisation's actual risk state may be worse than the formal picture suggests; the epistemic discipline of maintaining vigilance in the face of reassuring metrics and apparent good performance. Identified by Hopkins (2007) as a characteristic of mindful leaders. See also: Mindful Leadership.

Collective Mindfulness: The organisational practices and cultural norms that enable distributed information to be aggregated, challenged, and acted on at the level of the system rather than the individual component, the capacity for shared, active, ongoing awareness of the organisation's current operational state, including deviations from intended performance. Developed by Weick et al. (1999). See also: High-Reliability Organisation.

Cultural Filtering: The implicit or explicit discouragement of bad news within organisational hierarchies, producing the attenuation of safety-critical information through cultural rather than structural mechanisms. One of four mechanisms of Signal Attenuation identified in this book. See also: Signal Attenuation.

Deference to Expertise: The organisational practice of allowing decisions to migrate to the level of the hierarchy where the relevant expertise resides, regardless of formal seniority. One of the five processes of collective mindfulness identified by Weick and Sutcliffe (2001). See also: Collective Mindfulness, High-Reliability Organisation.

Drift into Failure: Dekker's (2011) account of how organisations migrate gradually toward the boundary of safe performance through the accumulation of individually reasonable decisions made under production pressure, without any single decision appearing unreasonable in its immediate context. See also: Production Gravity, Normalised Tolerance.

Due Diligence: The positive legal obligation imposed on officers of corporations under Australian work health and safety legislation and equivalent provisions in other jurisdictions to take reasonable steps to ensure that the organisation is meeting its safety obligations. The standard of due diligence is directly related to the concept of mindful leadership: officers who have not developed the practices required to maintain genuine awareness of their organisation's safety condition may not be discharging this obligation. See also: Mindful Leadership, Leadership Detachment.

Epistemic Honesty: The willingness to know what you do not know, to acknowledge the limitations of the picture that formal systems construct, and to actively seek the information that the picture is missing.

Developed in this book as the foundational quality of leadership that genuine active foresight requires. See also: Active Foresight, Leadership Detachment.

Ethical Fading: The process by which the moral dimensions of decisions become progressively less visible to the decision-makers who make them, as the decisions are framed in organisational terms, commercial, operational, or technical, that obscure their ethical content. Developed by Tenbrunsel and Messick (2004). See also: Moral Drift, Moral Disengagement.

High-Reliability Organisation (HRO): An organisation that operates a complex, high-hazard system with consistently lower rates of catastrophic failure than the technology and the operational environment would predict, through the sustained application of the five processes of collective mindfulness. Concept developed by La Porte and Consolini (1991) and elaborated by Weick et al. (1999). See also: Collective Mindfulness.

Incubation Period: Turner and Pidgeon's (1997) foundational concept: the pre-disaster phase in which the conditions for catastrophe develop gradually within the organisation, generating warning signals that are present within the system but are not correctly registered, interpreted, or acted upon. The incubation period is the phase addressed by the Blindspot Model. See also: Signal Attenuation, Turner, B.A.

Institutional Wilful Blindness: The organisational equivalent of wilful blindness, the condition in which an organisation has created, through the ordinary processes of its cultural and institutional life, a culture and information system that make it structurally unlikely that uncomfortable evidence will be fully assembled, accurately interpreted, and honestly communicated to the people who would have to act on it. Developed in this book, drawing on Sunstein and Vermeule (2001).

Just Culture: The cultural and institutional framework that allows safety-relevant information to be reported honestly without triggering the punitive responses that Signal Attenuation produces. Defined by Dekker (2007) as a culture that balances the need for accountability with the need for honest safety reporting. See also: Signal Attenuation.

Lag Indicator: A safety performance measure that captures the outcomes of past safety performance, injury rates, incident frequencies, and near-miss counts, rather than the developing conditions that will determine future performance. Contrasted with leading indicators. See also: Leading Indicator, Process Safety, Production Gravity.

Leading Indicator: A safety performance measure that reflects the current state of the safety management systems and conditions that will determine future safety performance, the integrity of barriers, the quality of safety culture, and the degree of procedural compliance, providing early warning of deterioration before that deterioration produces a serious incident. Contrasted with lag indicators. See also: Lag Indicator, Active Foresight.

Leadership Detachment: The fifth condition of the Blindspot Model: the structural loss of contact between senior leaders and operational reality that is a predictable consequence of organisational scale, hierarchy, and information filtering, producing a leadership team whose confidence in the organisation's safety performance is grounded not in accurate knowledge of the system's actual state but in the picture that the system's reporting structures have constructed for them. Developed in this book, drawing on Hopkins (2007). See also: Mindful Leadership, Blindspot Model.

Mindful Leadership: Hopkins' (2007) concept of leadership characterised by active, persistent, sceptical engagement with the gap between formal safety systems and operational reality, leaders who probe beneath the reassuring surface of formal reporting, visit operational environments, maintain direct engagement with frontline personnel, and treat the absence of bad news as a reason for increased vigilance rather than reassurance. See also: Chronic Unease, Leadership Detachment.

Moral Disengagement: Bandura's (1999) framework identifying eight psychological and social mechanisms through which individuals and organisations progressively disengage from the moral standards that would normally govern their behaviour, including moral justification, euphemistic labelling, displacement of responsibility, and diffusion of responsibility. See also: Moral Drift, Ethical Fading.

Moral Drift: The gradual erosion of ethical standards and moral attention that occurs through the same incremental, self-reinforcing mechanisms as Normalised Tolerance, the progressive revision of what counts as morally acceptable conduct through repeated exposure to morally compromised decisions without immediate accountability. Developed in this book, drawing on Bandura (1999) and Tenbrunsel and Messick (2004). See also: Normalised Tolerance, Ethical Fading.

Near-Miss Reclassification Cycle: The specific mechanism through which Normalised Tolerance operates: an event that is, objectively, a warning signal is investigated, a conclusion is reached that it is within acceptable limits, and the boundary of acceptable practice is imperceptibly extended. The next similar event is assessed against the revised boundary. Over time, the organisation's understanding of acceptable operational practice has been fundamentally altered without any explicit decision to alter it. Developed in this book. See also: Normalised Tolerance.

Normalisation of Deviance: Vaughan's (1996) concept describing the process by which a deviation from established safe practice, through its repeated occurrence without catastrophic outcome, is gradually reclassified within the organisational culture as an acceptable characteristic of normal operations rather than a hazard requiring urgent attention. See also: Normalised Tolerance.

Normalised Tolerance: The second condition of the Blindspot Model: the gradual, culturally embedded process through which repeated exposure to a risk that does not immediately produce harm generates a revised, more permissive understanding of what constitutes acceptable operational practice. Developed in this book, incorporating and extending Vaughan's (1996) normalisation of deviance concept. See also: Blindspot Model, Normalisation of Deviance.

Perfect Blindspot: The organisational state in which all five conditions of the Blindspot Model, Signal Attenuation, Normalised Tolerance, Structural Secrecy, Production Gravity, and Leadership Detachment, are operating simultaneously, with no single condition providing any compensating protection against the blindness produced by the others. Developed in this book. See also: Blindspot Model.

Personal Safety: The management of risks to individual workers from the ordinary hazards of their work, slips, trips, manual handling injuries, and similar events, typically measured through lost-time injury rates and recordable incident rates. Distinguished from process safety. See also: Process Safety.

Process Safety: The management of the major hazard risks inherent in the processing of flammable, toxic, or explosive materials, and the risks of fires, explosions, and toxic releases that have the potential to cause catastrophic harm. Distinguished from personal safety. Hopkins (2008) documented the dangerous confusion of these two categories at BP Texas City. See also: Personal Safety.

Production Gravity: The fourth condition of the Blindspot Model: the constant economic and operational pressure that pulls organisational decision-making toward the performance of productive work and away from the maintenance of safety margins, producing incremental, individually justified erosions of those margins that, in combination, leave the organisation without adequate protection against catastrophic failure. Developed in this book, drawing on Dekker (2011) and Hopkins (2008). See also: Blindspot Model, Drift into Failure.

Psychological Safety: Edmondson's (2018) concept: the shared belief that the team environment is safe for interpersonal risk-taking, specifically including the raising of safety concerns, the cultural condition that most strongly predicts effective safety communication across hierarchical boundaries. See also: Signal Attenuation, Just Culture.

Reasonably Practicable: The legal standard applied in Australian work health and safety legislation to determine what measures a person conducting a business or undertaking must take to ensure safety. A measure is reasonably practicable if it is reasonably able to be done, taking into account and weighing up all relevant matters, including the likelihood of the hazard occurring, the degree of harm, what the person knows or ought reasonably to know, and the cost of the measure.

Safety Intelligence Function: A dedicated organisational function charged with assembling and maintaining the complete major hazard risk picture across all five conditions of the Blindspot Model, analysing trends for evidence of developing blindspot conditions, and providing senior leadership and the board with integrated risk intelligence. Distinguished from the safety management function by its prospective orientation. Developed in this book. See also: Active Foresight.

Signal Attenuation: The first condition of the Blindspot Model: the systematic filtering, softening, and loss of safety-critical information as it moves upward through organisational hierarchies, such that warning signs present at the operational level fail to reach the decision-makers with the authority and responsibility to act on them. Developed in this book, drawing on Turner (1997), Hopkins (2019), and Weick (1995). See also: Blindspot Model.

Structural Secrecy: The third condition of the Blindspot Model: the fragmentation of safety-critical information across organisational boundaries, functional silos, geographic sites, and hierarchical layers, such that no individual or team has access to the complete picture of accumulating risk. The term was originally developed by Vaughan (1996) and extended in this book as a condition of the Blindspot Model. See also: Blindspot Model, Vaughan, D.

Swiss Cheese Model: Reason's (1997) model of organisational accidents, in which each layer of defence against failure is represented as a slice of Swiss cheese, effective in most places but with gaps. An accident occurs when the gaps in successive layers momentarily align, allowing a trajectory of accidental opportunity to pass through all layers simultaneously. See also: Reason, J.

Threat and Error Management (TEM): A framework developed by Helmreich and colleagues at the University of Texas Human Factors Research Project for understanding and managing the safety of flight operations by identifying, in real time, the threats present in the operational environment and the errors being made in response to those threats, before they compound into incidents or accidents. See also: Active Foresight.

Wilful Blindness: The deliberate avoidance of knowledge that would impose unwelcome obligations, a form of culpable ignorance for which the ignorant party bears moral and, in some jurisdictions, legal responsibility. In the organisational disaster context, wilful blindness typically operates not through conscious individual choice but through the creation of institutional cultures and information systems that make the honest assembly and communication of uncomfortable evidence structurally unlikely. See also: Institutional Wilful Blindness.

Consolidated Bibliography

Master References

All sources cited in this book are listed below in alphabetical order by author surname. Sources are formatted according to the Harvard referencing style. All sources are verifiable and, where publication dates are noted, were published within the 30 years preceding the writing of this book, with the exception of foundational works cited for their historical significance.

A

Argyris, C. (1990) *Overcoming Organizational Defenses: Facilitating Organizational Learning*. Boston: Allyn and Bacon.

Australian Broadcasting Corporation (2016) *'The Wittenoom Story: Australia's Greatest Industrial Disaster', Four Corners*. Sydney: ABC.

B

Baker, J.A. (2007) *The Report of the BP US Refineries Independent Safety Review Panel*. Washington: BP.

Bandura, A. (1999) 'Moral disengagement in the perpetration of inhumanities', *Personality and Social Psychology Review*, 3(3), pp. 193–209.

Bazerman, M.H. and Tenbrunsel, A.E. (2011) *Blind Spots: Why We Fail to Do What's Right and What to Do About It*. Princeton: Princeton University Press.

Bly, M. (2010) *Deepwater Horizon Accident Investigation Report*. London: BP plc.

C

Columbia Accident Investigation Board (2003) *Columbia Accident Investigation Board Report*. Volume 1. Washington: NASA.

Cullen, L.W.D. (1990) *The Public Inquiry into the Piper Alpha Disaster*. London: HMSO.

D

Dekker, S. (2002) *The Field Guide to Human Error Investigations*. Aldershot: Ashgate.

Dekker, S. (2007) *Just Culture: Balancing Safety and Accountability*. Aldershot: Ashgate.

Dekker, S. (2011) *Drift into Failure: From Hunting Broken Components to Understanding Complex Systems*. Farnham: Ashgate.

Dekker, S. (2014) *The Field Guide to Understanding Human Error*. 3rd edn. Farnham: Ashgate.

E

Edmondson, A.C. (2018) *The Fearless Organization: Creating Psychological Safety in the Workplace for Learning, Innovation, and Growth*. Hoboken: Wiley.

F

Flin, R., O'Connor, P. and Crichton, M. (2008) *Safety at the Sharp End: A Guide to Non-Technical Skills*. Aldershot: Ashgate.

French, P.A. (1979) 'The corporation as a moral person', *American Philosophical Quarterly*, 16(3), pp. 207–215.

G

Gawande, A. (2009) *The Checklist Manifesto: How to Get Things Right*. New York: Metropolitan Books.

Grenfell Tower Inquiry (2019) *Phase 1 Report: Overview*. London: HMSO.

Grenfell Tower Inquiry (2024) *Phase 2 Report*. London: HMSO.

H

Helmreich, R.L., Merritt, A.C. and Wilhelm, J.A. (1999) 'The evolution of Crew Resource Management training in commercial aviation', *International Journal of Aviation Psychology*, 9(1), pp. 19–32.

Higginbotham, A. (2019) *Midnight in Chernobyl: The Untold Story of the World's Greatest Nuclear Disaster*. London: Penguin.

Higginbotham, A. (2023) *Challenger: A True Story of Heroism and Disaster on the Edge of Space*. London: Penguin Viking.

Hollnagel, E., Wears, R.L. and Braithwaite, J. (2015) *From Safety-I to Safety-II: A White Paper*. Published by the University of Southern Denmark, University of Florida, USA, and Macquarie University, Australia.

Hollnagel, E., Woods, D.D. and Leveson, N. (eds.) (2006) *Resilience Engineering: Concepts and Precepts*. Aldershot: Ashgate.

Hopkins, A. (2000) *Lessons from Longford: The Esso Gas Plant Explosion*. Sydney: CCH Australia.

Hopkins, A. (2007) *Lessons from Gretley: Mindful Leadership and the Law*. Sydney: CCH Australia.

Hopkins, A. (2008) *Failure to Learn: The BP Texas City Refinery Disaster*. Sydney: CCH Australia.

Hopkins, A. (2012) *Disastrous Decisions: The Human and Organisational Causes of the Gulf of Mexico Blowout*. Sydney: CCH Australia.

Hopkins, A. (2014) *Learning from High Reliability Organisations*. Sydney: CCH Australia.

Hopkins, A. (2019) *Organising for Safety: How Structure Creates Culture*. Sydney: CCH Australia.

I

Institute of Medicine (1999) *To Err Is Human: Building a Safer Health System*. Washington: National Academy Press.

K

Kahneman, D. (2011) *Thinking, Fast and Slow*. New York: Farrar, Straus and Giroux.

Klein, G. (1998) *Sources of Power: How People Make Decisions*. Cambridge: MIT Press.

L

La Porte, T.R. and Consolini, P.M. (1991) 'Working in practice but not in theory: theoretical challenges of high-reliability organizations', *Journal of Public Administration Research and Theory*, 1(1), pp. 19–47.

Leonard, M., Graham, S. and Bonacum, D. (2004) 'The human factor: the critical importance of effective teamwork and communication in providing safe care', *Quality and Safety in Health Care*, 13(Suppl 1), pp. 85–90.

M

Macfie, R. (2013) *Tragedy at Pike River Mine: How and Why 29 Men Died*. Wellington: Awa Press.

Morrison, E.W. and Milliken, F.J. (2000) 'Organizational silence: a barrier to change and development in a pluralistic world', *Academy of Management Review*, 25(4), pp. 706–725.

Muller, J.Z. (2018) *The Tyranny of Metrics*. Princeton: Princeton University Press.

N

NASA (2023) *Aviation Safety Reporting System: Program Overview*. Moffett Field: NASA Ames Research Center.

National Commission on the BP Deepwater Horizon Oil Spill and Offshore Drilling (2011) *Deep Water: The Gulf Oil Disaster and the Future of Offshore Drilling*. Washington: US Government Printing Office.

P

Perrow, C. (1999) *Normal Accidents: Living with High-Risk Technologies*. 2nd edn. Princeton: Princeton University Press.

Pidgeon, N.F. and O'Leary, M. (2000) 'Man-made disasters: why technology and organizations sometimes fail', *Safety Science*, 34(1-3), pp. 15–30.

Plokhy, S. (2018) *Chernobyl: History of a Tragedy*. London: Allen Lane.

Presidential Commission on the Space Shuttle Challenger Accident (1986) *Report of the Presidential Commission on the Space Shuttle Challenger Accident*. Washington: US Government Printing Office.

R

Reason, J. (1997) *Managing the Risks of Organizational Accidents*. Aldershot: Ashgate.

Reason, J. (2000) 'Human error: models and management', *British Medical Journal*, 320(7237), pp. 768–770.

Rees, J.V. (1994) *Hostages of Each Other: The Transformation of Nuclear Safety since Three Mile Island*. Chicago: University of Chicago Press.

Royal Commission on the Pike River Coal Mine Tragedy (2012) *Final Report*. Volumes 1 and 2. Wellington: New Zealand Government.

S

Sheen, J.B. (1987) *MV Herald of Free Enterprise: Report of Court No. 8074 Formal Investigation*. London: HMSO.

Singer, J. (2022) *There Are No Accidents: The Deadly Rise of Injury and Disaster*. New York: Simon and Schuster.

Snook, S.A. (2000) *Friendly Fire: The Accidental Shootdown of US Black Hawks over Northern Iraq*. Princeton: Princeton University Press.

Sunstein, C.R. and Vermeule, A. (2001) 'Interpretation and institutions', *Michigan Law Review*, 101(4), pp. 885–951.

Sylvestre, C. (2020) *Third Generation Safety: The Missing Piece*. Safety Institute of Australia.

T

Tenbrunsel, A.E. and Messick, D.M. (2004) 'Ethical fading: the role of self-deception in unethical behavior', *Social Justice Research*, 17(2), pp. 223–236.

Turner, B.A. and Pidgeon, N.F. (1997) *Man-Made Disasters*. 2nd edn. Oxford: Butterworth-Heinemann.

Tversky, A. and Kahneman, D. (1974) 'Judgment under uncertainty: heuristics and biases', *Science*, 185(4157), pp. 1124–1131.

U

US Chemical Safety and Hazard Investigation Board (2007) *Investigation Report: Refinery Explosion and Fire, BP Texas City, Texas*. Washington: CSB.

V

Vaughan, D. (1996) *The Challenger Launch Decision: Risky Technology, Culture, and Deviance at NASA*. Chicago: University of Chicago Press.

W

Weick, K.E. (1995) *Sensemaking in Organizations*. Thousand Oaks: Sage.
Weick, K.E. and Sutcliffe, K.M. (2001) *Managing the Unexpected: Assuring High Performance in an Age of Complexity*. San Francisco: Jossey-Bass.
Weick, K.E., Sutcliffe, K.M. and Obstfeld, D. (1999) 'Organizing for high reliability: processes of collective mindfulness', *Research in Organizational Behavior*, 21, pp. 81–123.
WorkSafe Victoria (1999) *Esso Longford Gas Plant Accident: Report of the Longford Royal Commission*. Melbourne: Victorian Government.

Index

Note: *Italic* page numbers refer to figures.

Major organisational disasters are not random. They are predictable. In the months and years before every catastrophic failure, before the Deepwater Horizon blowout, the Grenfell Tower fire, the Space Shuttle disasters, the Pike River explosion, the warning signs were already present. They were systematically unseen. Organisational blindness to accumulating risk is not a malfunction. In many organisations, it is the system working exactly as designed. This book explains why and what can be done about it.

The Blindspot Before the Disaster identifies the five organisational conditions that converge to produce this failure of sight: Signal Attenuation, Normalised Tolerance, Structural Secrecy, Production Gravity, and Leadership Detachment. Together they form the Blindspot Model, a practical diagnostic framework grounded in four decades of disaster scholarship and demonstrated across 11 of the most significant industrial catastrophes of the past 40 years.

This book is written for the leaders, safety professionals, and governance practitioners who are responsible for preventing harm before it occurs. It bridges the gap between rigorous academic disaster theory and the practical needs of people working in high-consequence environments. It does not assign blame. It offers something more useful: a structured, evidence-based way of seeing the danger that is already in the room.

Tom Bourne is an Australian author, workplace safety educator, speaker, and leadership commentator with more than 25 years of experience across the mining and resources, government, and professional education sectors. He has published more than 16 books on workplace safety, leadership, and organisational accountability. He is based in Perth, Western Australia.

Routledge
Taylor & Francis Group
www.routledge.com

Routledge titles are available as eBook editions in a range of digital formats

an informa business

ISBN 978-1-04-139337-5

9 781041 393375